After Euclid

CSLI Lecture Notes
Number 175

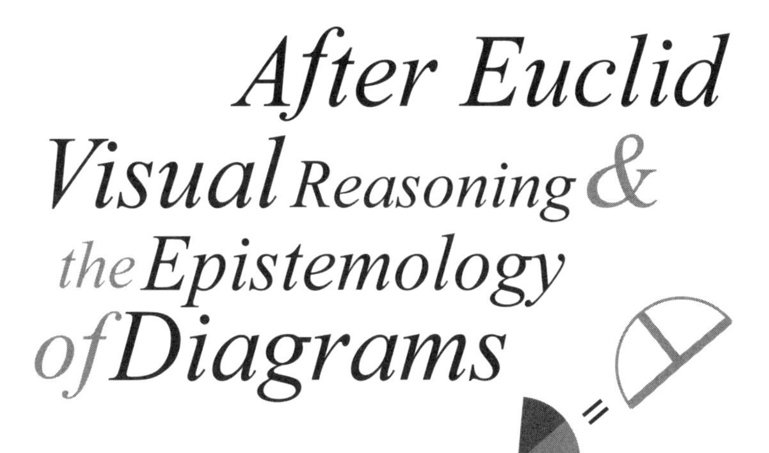

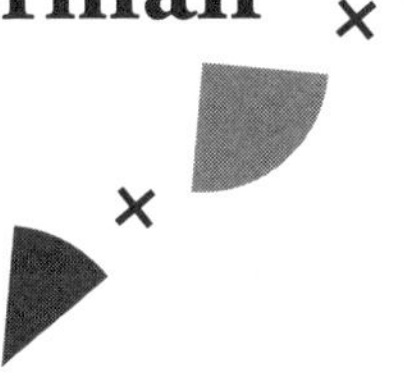

Center for the Study of Language and Information
Leland Stanford Junior University
Printed in the United States
10 09 08 07 06 1 2 3 4 5

Library of Congress Cataloging-in-Publication Data

Norman, Jesse.
After Euclid : visual reasoning and the epistemology of diagrams /
by Jesse Norman.

p. cm. – (CSLI lecture notes ; no. 175)

Includes bibliographical references (p.) and index.
ISBN 1-57586-510-6 (pbk. : alk. paper)
ISBN 1-57586-509-2 (cloth cover : alk. paper)

1. Logic diagrams. 2. Geometry. I. Title. II. Series.

BC136.N67 2006
160–dc22 2005031849
CIP

∞ The acid-free paper used in this book meets the minimum requirements of the American National Standard for Information Sciences—Permanence of Paper for Printed Library Materials, ANSI Z39.48-1984.

CSLI was founded in 1983 by researchers from Stanford University, SRI International, and Xerox PARC to further the research and development of integrated theories of language, information, and computation. CSLI headquarters and CSLI Publications are located on the campus of Stanford University.

CSLI Publications reports new developments in the study of language, information, and computation. Please visit our web site at
http://cslipublications.stanford.edu/
for comments on this and other titles, as well as for changes and corrections by the author and publisher.

"In these days the angel of topology and the devil of abstract algebra fight for the soul of every individual discipline of mathematics."

Hermann Weyl, *The Continuum*

To My Mother and Father

Contents

Preface

I came to Euclid in the same way I imagine most people have over the centuries: constructing geometrical diagrams at school with a compass and ruler, happy when I could make the lines join up at roughly the right points, rarely if ever reasoning through the arguments, and quite unaware that I was entering a world of ideas well over two thousand years old.

When I came back to think *about* Euclid, however, it was via a route that even philosophers might consider circuitous. I had been introduced by Lee Auspitz to the work of two of the least-read and most interesting of modern philosophers: Michael Oakeshott and C.S. Peirce. In *Experience and Its Modes* (1933) Oakeshott advances a view of human experience as organized into different 'modes' or categories, each mode itself structured by a distinct set of leading concepts and presuppositions. Oakeshott theorized three such modes in that book—those of science, history and 'practice'—and to these he later added a fourth, the mode of 'poetry' or artistic experience. It is a picture inspired by Hegel and also crucially, as Auspitz has since argued, by Spinoza.

Whatever the weaknesses of this view—and Oakeshott himself came to revise it in some important respects—it seemed to me to have among its merits two particular ones worthy of note. First, it recognized the remarkable diversity and apparent heterogeneity of human experience, including the experience of thinking; and secondly, it thereby directed philosophical attention to the task of picking apart and describing the distinctive concepts and principles governing each candidate mode. It was surely a threshold condition on something's being a mode at all that it be capable of being analyzed in this way.

Among other things, this book is an attempt to discharge the major part of that task for a small but well-known type of basic geometrical reasoning. It does so in the course of pursuing an epistemological investigation into how, if at all, such reasoning can justify belief and confer knowledge; and this, of course, is its central purpose. As such, it is logically prior to the broader picture sketched above, and in no way draws on Oakeshottian assumptions that, however beguiling they are, may rightly be regarded as philosophically much-contested ground.

However, the book also falls well short of a fully modal analysis 'at the other end', so to speak. This would extend into a deeper consideration of the target reasoning, understood not only conceptually but semiotically, that is in terms of the different sign-relations involved. As such, it might draw on Peirce's writings on the nature of signs, and in particular on his late attempts at a comprehensive taxonomy of signs. Such an extended analysis is, again, deliberately not part of the focus of the book, which is firmly epistemological. This is perhaps just as well given the controversial standing of semiotics among many philosophers, and the many different causes in which that term has been enlisted, rightly or wrongly; in recent years. But I mention it so that reader can understand the broader project involved. As should already be clear, this project originates in Lee Auspitz's own work, and to him and it I owe profound thanks.

This book was conceived and written at the Philosophy Department of University College London, originally under a British Academy (AHRB) postgraduate award and then under the first *MIND* Studentship. I am most grateful to the Department, to the AHRB and to the *MIND* Association for their support.

Parts of this and/or related work have been presented in various forms at AHRB seminars in London and Oxford; at conferences in Dubrovnik, Liverpool and Cambridge; and at the Annual Meeting of the C.S. Peirce Society, American Philosophical Association, Eastern Division. I am grateful to the participants at these sessions for their questions and comments.

I would especially like to thank my doctoral supervisors Marcus Giaquinto, Mark Kalderon and Mike Martin. In particular, I have learned an enormous amount from the painstaking criticism and thought-provoking ideas of Marcus Giaquinto, my principal supervisor. I am also very grateful to my examiners, Peter Clark and David Galloway, for their comments on the original thesis, and for their other assistance.

Finally, I would like to thank David Bostock, Steve Butterfill, Tim Crane, Tom Crowther, Steve Gross, John Lucas, Robert A.G. Monks, Matthew Nudds, Hanna Pickard, Matthew Soteriou, Barry C. Smith, Maya Spener, Scott Sturgeon, Jerry Valberg and Jo Wolff for their ideas, their comments and their support. Ken Manders and Dennis Potter generously shared unpublished work of theirs on diagrams. The late Jon Barwise kindly encouraged me to study Euclid's diagrams over a memorable lunch with Gerry Allwein in Bloomington. Ian Page of the Classical Opera Company introduced me to the Mozart quotation in the final chapter.

This book could not have been written without the interruptions of Kate Bingham and Sam, Nell and Noah Norman. I dedicate it to my parents, Torquil and Anne Norman.

Readers who have any questions, comments or corrections are invited to contact the author by email at jesse.norman@dial.pipex.com, or through www.jessenorman.com.

1

An Old Kind of Reasoning

1.1 Introduction

Can reasoning with diagrams be epistemically valuable? Can it confer justification, or knowledge? A common view holds that the role of diagrams in mathematical arguments is merely heuristic. On this view, diagrams serve as illustrations, whose function is to make an argument given wholly in words or formulas easier for a reasoner to grasp. The text of the argument alone confers justification, however, and so the diagram is—whatever its psychological merits—epistemically redundant. On an older view, diagrams have epistemic value, but this value is understood via an appeal, now widely considered philosophically dubious, to a postulated special faculty of 'intuition'. Such a faculty is 'special' in that it does not merely draw on the perceptual and ratiocinative faculties required to follow a given presentation of an argument in words or formulas; it is supposed to be an independent and fundamental source of mathematical justification. On this view, diagrams can play a non-redundant role in conferring justification or knowledge, but they do so by utilizing a reasoner's intuition.

To many these have seemed—within the traditional picture of mathematical knowledge as *a priori*—to exhaust the available alternatives, and worries about the existence or epistemic status of intuition have in turn motivated the dismissal of diagrams. Thus, on a standard story, one of the goals of 19th Century mathematics was to supersede appeals to intuition as a ground for knowledge, with Euclid's geometry—in which diagrams are ubiquitous—an important target. On this interpretation, Euclid's presentation is insufficient to justify belief or confer knowledge in Euclidean geometry. It was only with the work of Hilbert that the possibility of a fully rigorous presentation of Euclidean geometry became clear, and such a presentation makes, it is claimed, no non-redundant use of diagrams. This claim was made by Hilbert himself in a lecture of 1894:

> A system of points, lines and planes is called a diagram or figure [*Figur*]. *The proof* [of the theorem Hilbert is discussing] *can indeed be given by calling on a suitable figure, but this appeal is not at all necessary.* [It merely] makes the interpretation easier [*erleichtert die Auffassung*], and it is a fruitful means of discovering new propositions. Nevertheless, care, since it can easily be misleading. *A theorem is only proved when the proof is completely independent of the diagram.*[1]

1.2 A Fallacy in Euclid's Geometry?

The denial that reasoning with diagrams can have epistemic value has often been supported by the claim that diagrams (and other visual representations) are not reliable, or are actively misleading. A flat map of the world does not preserve correct information about distances: someone wishing to fly by the shortest route from London to San Francisco would be ill-advised to follow a straight line on an airline map, for the shortest route between these cities is not this line, but the shorter of the two arcs of the great circle on which they lie. And within mathematical research, the discovery in the last century of counterintuitive results in analysis—such as space-filling curves and continuous but nowhere differentiable functions—had the effect of undermining the perceived reliability of diagrams.[2]

Let's consider a specific case of the apparently unreliable and misleading nature of diagrams in Euclid's geometry: a well-known fallacious argument to the effect that all triangles are isosceles. I reproduce the argument below, as given by E. A. Maxwell.[3] However, for later convenience I have also annotated the various steps of the argument in square brackets on the right hand side so as to bring out its structure. The reader may wish to cover over these annotations in following the argument for the first time here.

---

[1] Quoted in Hallett 1994, non-German emphasis added. Note what Hallett calls the 'rather abstract characterization' of the figure.

[2] Hahn 1933.

[3] Maxwell 1959, Ch. 2. Note that '≡' denotes congruence. 'SAS' (side-angle-side), 'ASA' (angle-side-angle) and 'Rt. ∠-H-S' (right angle-hypotenuse-side) denote established rules for congruence of triangles in Euclid. The argument appears to have been invented by Rouse Ball, in Rouse Ball 1905, pp. 44-45.

To prove that every triangle is isosceles.

Given: A triangle ABC.

Required: To prove that, necessarily, AB = AC.

Construction: Let the internal bisector of the angle A meet the perpendicular bisector of BC [from D] at O. Draw OD, OQ, OR perpendicular to BC, CA, AB respectively.

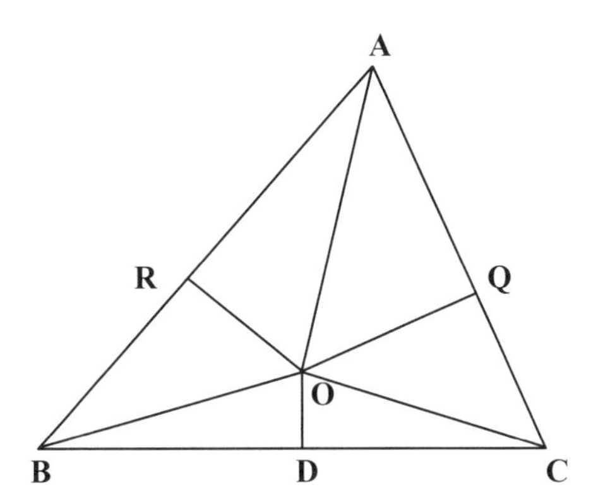

Proof:

(1)	DO = DO		[self-identity of DO]
(2)	DB = DC		[bisection of BC]
(3)	∠ODB = ∠ODC		[OD perp. to BC]
(4)	ΔODB ≡ ΔODC	(SAS)	[1, 2, 3]
(5)	OB = OC		[4]

Also

(6)	AO = AO		[self-identity of AO]
(7)	≡ RAO = ≡ QAO		[bisection of ≡ BAC]
(8)	≡ ARO = ≡ AQO		[OR perp. to AB, OQ to AC]
(9)	ΔARO ≡ ΔAQO	(ASA)	[6, 7, 8]
(10)	AR = AQ		[9]
(11)	OR = OQ		[9]

Hence in triangles OBR, OCQ,

(12) $\angle ORB = \angle OQC$ = right angle [OR perp. to AB, OQ to AC]
(13) $OB = OC$ (proved) [5]
(14) $OR = OQ$ (proved) [11]
(15) $\Delta ORB \equiv \Delta OQC$ (rt. $\equiv$-H-S) [12, 13, 14]
(16) $RB = QC$ [15]

Finally,

(17) $AB = AR + RB$ [by inspection of AB]
(18) $AB = AQ + QC$ (proved) [10, 16, 17]
(19) $AB = AC$ [18; by inspection of AC]

Spelling out the inferential structure of the argument in this way makes explicit that, with two exceptions, all the lines of the argument are warranted either by already-established results (e.g. line 13), or by established rules of logic (e.g. line 1, by the rule that everything is self-identical), or by the text of the construction rubric. As an example of the latter, the claim that DB = DC (line 2) is warranted, not by the visual appearance of the diagram, but because the reasoner has already been instructed to draw OD as the bisector of BC.

The two exceptions are lines 17 and 19. These claims are not warranted by prior results, by established rules of logic or by the text of the (admittedly over-concise) construction rubric. Rather, they are supposedly warranted, wholly or partly, by the reasoner's reading claims off directly from the diagram. Moreover, for reasons that will shortly become evident, it is here that the argument goes wrong. So the case that the diagram is misleading here is clear-cut: it seems quite obvious from the diagram that AB = AR + RB, for example, but someone who thought she could read this claim off from the diagram in this way would have gone wrong.

If the charge of being misleading is to go through, however, we also need a reason to believe that the fault is, so to speak, on Euclid's side. After all, we would not claim that airline maps were misleading if they came complete with a set of explicit instructions as to how they were to be properly read; we would say that the person who read the map wrongly without having followed the instructions had simply made a mistake of her own, a mistake which could have been avoided had she been sufficiently careful. The same is true of someone with a tacit mastery of maps, in the absence of explicit instructions: we would not count an airline map misleading if such a person, knowing it was an airline map, misread it.

Is the diagram here misleading if we apply this stricter standard? The first thing to say is that even someone moderately familiar with Euclid will ask whether there is not more than one case to be considered here. Thus Greenberg's presentation of the same argument considers three possible cases: one where point O—the intersection of the bisector of angle A and the perpendicular bisector of BC—lies inside triangle ABC, one where it lies on line BC, and one where it lies outside triangle ABC.[4] The first of these has already been discussed; the other cases are diagrammed below:

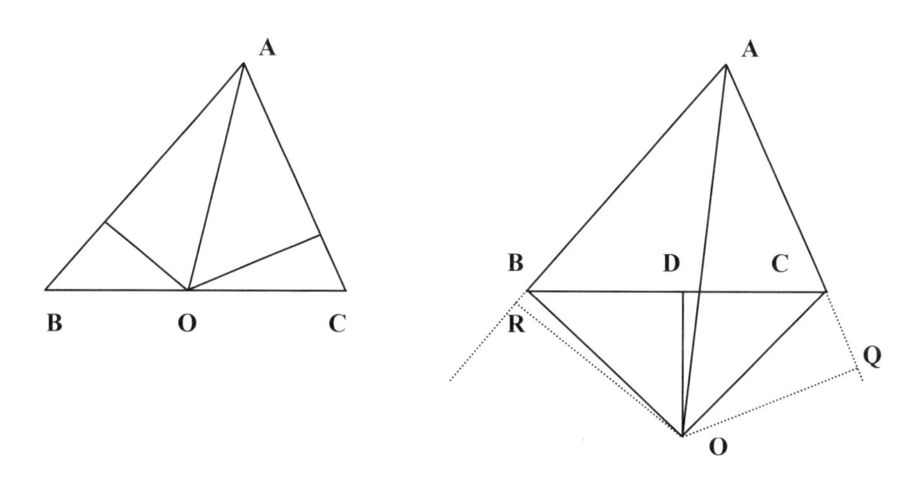

As Greenberg shows, a very similar argument to that given above goes through for the case represented by the left hand diagram. It also goes through for that represented by the right hand diagram above, except that the conclusion is reached by subtraction, not addition, of line segments in lines 17 and 18. At this point the prospect is bleak for Euclid, since it now looks as though the fallacious argument succeeds by exhaustion. Merely reading off lines 17 and 19 from the earlier diagram was non-conclusive; it left open alternative possibilities for the location of the point O, for which the argument might not work. But the alternatives above have now been closed off. So now it appears that unless the diagram is misleading the reasoner, Euclid's argument goes through, and the fallacious conclusion that all triangles are isosceles is established.

There is, however—if we continue to talk of cases—a fourth case to be considered. This can be represented diagrammatically as follows:

---

[4] Greenberg 1993, pp. 24-5.

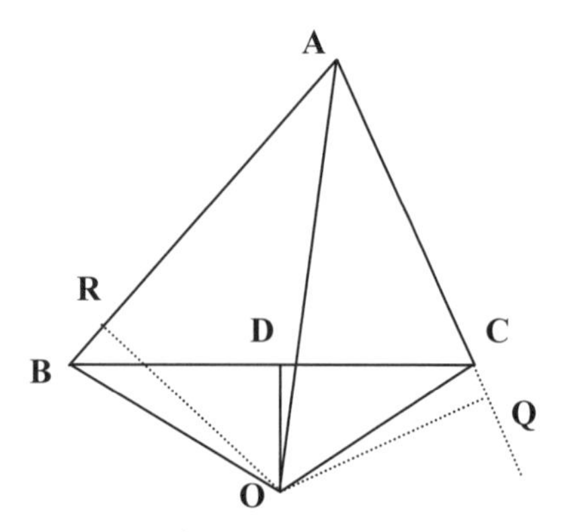

The argument above does not go through for the case represented by this diagram. For though RB and QC are equal, it does not follow that AB and AC are equal; although R lies between A and B, Q does not lie between A and C. This is the result Euclid needs, for here it is true both that the diagram is not misleading and that the argument is blocked. Moreover, it is a result that, plausibly, someone with only moderate expertise in Euclid's geometry ought to be able to reach. In the first place, it is suggested by the third case already considered above, where it is evident that closing the gap between D and O a little would have the effect of making R fall between A and B, with Q still falling outside AC. But secondly, although I have been following normal practice in talking of different cases here—and although other arguments in Euclid do require analysis by cases—in fact there are no different cases to be considered: the situation represented by the diagram above—in which R falls between AB and Q falls outside AC—is the *only* correctly drawn one. The other diagrams above are misdrawn, as careful examination should make clear to the reader.

But now we can say this: this and other fallacious arguments are often advanced as evidence of the misleading nature of Euclid's diagrams. But here at least, the argument only appears to succeed because the correct diagram is never included in the presentation of the argument. None of Greenberg, Maxwell or Rouse Ball includes it, for example; rather, each presentation utilizes—and relies for its plausibility on—various incorrectly drawn diagrams. But someone who accepts an incorrectly drawn diagram provided in a given presentation of an argument has made a mistake; the fault is not Euclid's. Indeed, not only is the correct diagram not misleading; it makes it evident where the error in the argument falls. It can be appropriate to include misdrawn diagrams as an educational device for students. But it is wrong to infer from this clear and correctable misrepresentation that diagrams in Euclid are generally misleading. Although diagrams in Euclid may and do sometimes need careful handling, this objection in itself offers no reason to think either that they are misleading to a suitably competent practitioner, or that reasoning with diagrams of the kind(s) we find in Euclid is generally fallacious. We might—and this brings out the wider point—make similar remarks about reasoning with quantifiers in a logical language, for example, or reasoning with expressions for negation in a natural language.

1.3 Diagrams and Intuition

So what, then, of the other view I mentioned above, that reasoning with diagrams can have epistemic value? Must someone who holds this view, and who is an apriorist about mathematical justification, postulate a faculty of intuition in order to explain why? The philosopher who is perhaps most closely associated with positive claims for a faculty of intuition is Kant, and the influence of Euclid's geometry on the Critique of Pure Reason is well known. In the 'B' Preface, Kant describes the main goal of his work as an 'an attempt to transform the accepted procedure of metaphysics, undertaking an entire revolution according to the example of the geometers and natural scientists.' In the Transcendental Aesthetic he takes the status of geometry as a synthetic *a priori* description of space to be an 'apodeictic certainty'. And in the Doctrine of Method, he gives a worked example of someone following an argument in Euclid to illustrate his doctrine that intuition of a diagram or figure is required for geometrical knowledge.[5]

Kant's belief that Euclid's geometry is the science of space is, of course, widely held to be untenable. But many commentators have also been dismissive of his claims about geometrical reasoning. The discovery of logical gaps in Euclid, many of them traceable to the lack of axioms giving an explicit theory of order for points in the line, has served to undermine Euclid's claim to rigor. And a further worry is that Kant, while doubtless familiar with the mathematics issuing from Descartes' *Géometrie*, did not foresee the degree to which the later development of analytic geometry would undermine the view of intuition described above. In analytic geometry, the plane is defined as the set of ordered pairs of real numbers, and straight lines as subsets of all pairs $<x, y>$ satisfying equations of the form $ax + by + c = 0$ (with a and b not both equal to 0). This allows geometrical properties to be translated into algebraic properties of real variables. On a standard modern presentation, the reals are not understood in spatial terms,

---

[5] Bxxii; A716/B744; A47/B64ff.

however, and so—at least on some views—geometry ceases to be an independent discipline and becomes a branch of real analysis.[6] But real analysis does not employ diagrams—the thought continues—at least in any epistemically non-redundant way; and there does not seem to be any place here for geometrical reasoning of the sort that Kant apparently regarded as exemplary of the synthetic *a priori*. So Kant is mistaken in thinking that intuition of a diagram is required for geometrical knowledge; and we can plausibly attribute the source of his mistake to his lack of a space-independent understanding of real numbers.

The case against both Kant and Euclid was eloquently made by Bertrand Russell in a series of writings at the beginning of the 20th Century. Perhaps under the influence of his reading of Leibniz, Russell rejects any epistemic role for diagrams in geometry:

> Formerly, it was held by philosophers and mathematicians alike that the proofs in Geometry depended on the figure; nowadays, this is known to be false. In the best books there are no figures at all. The reasoning proceeds by the strict rules of formal logic from a set of axioms laid down to begin with. If a figure is used, all sorts of things seem obviously to follow, which no formal reasoning can prove from the explicit axioms, and which, as a matter of fact, are only accepted because they are obvious. By banishing the figure, it becomes possible to discover *all* the axioms that are needed; and in this way all sorts of possibilities, which would otherwise have remained undetected, are brought to light.[7]

Kant, again, bears much of the responsibility for this error, according to Russell, through his lack of logical sophistication and consequent unnecessary and erroneous emphasis on intuition:

> Kant, having observed that the geometers of his day could not prove their theorems by unaided arguments, but required an appeal to the figure, invented a theory of mathematical reasoning according to which the inference is never strictly logical, but always requires the support of what is called intuition.[8]

Russell does not shrink from drawing the conclusion that, since Euclid's arguments employ diagrams, they do not justify; and indeed that

[6] On the translation of geometry into analysis see, for example, Hartshorne 2000a and 2000b.

[7] Russell 1901, p. 93.

[8] Russell 1919, p. 145 (cf. the very similar remarks at Russell 1901, p. 96).

perhaps none of the arguments advanced before at least the nineteenth century was deductively valid.

> It is perfectly true, for example, that anyone who attempts, without the use of the figure, to deduce Euclid's seventh proposition from Euclid's axioms, will find the task impossible; and there probably did not exist, in the eighteenth century, any single logically correct piece of mathematical reasoning, that is to say, any reasoning which correctly deduced its result from the explicit premisses laid down by the author. Since the correctness of the result seemed indubitable, it was natural to suppose that mathematical proof is something different from logical proof. But the fact is, that the whole difference lay in the fact that mathematical proofs were simply unsound.[9]

On Russell's view, then, 'proofs' containing diagrams are to be replaced by proofs containing only sentences, claims advanced on the basis of intuition are to be replaced by claims advanced on the basis of logic, and Kant's authority and doctrines, in this regard at least, are pernicious. It is a short step to diagnose Kant's error as lying in the weakness of the then-available logic, a weakness by which Kant was, unlike Leibniz, supposedly untroubled.

Similar views have been expressed in extreme form by Alberto Coffa, who identifies the start of what he considers the modern semantic tradition with 'Conceptualism', described as follows:

> Conceptualism is defined by an enemy, a goal, and a strategy: the enemy was Kant, the goal was the elimination of pure intuition from scientific knowledge, and the strategy was the creation of semantics as an independent discipline...
>
> Bolzano's problem [i.e. to prove that a continuous real function that takes values above and below zero, must also take a zero value somewhere in between] looks like a problem only to someone who [sc. unlike Kant and his followers] has already understood that intuition is not an indispensable aid to mathematical knowledge, but rather a cancer that has to be extirpated in order to make mathematical progress possible.[10]

That is, for Coffa what Bolzano is rejecting in a pioneering way is the attempt to argue for this or any mathematical claim by appealing to a figure

[9] Russell 1903, p. 457.

[10] Coffa 1982, pp. 679, 686.

or diagram, in relation to which the reasoner is supposed to exercise some kind of faculty of intuition.

The views expressed by Coffa and Russell have now become fairly orthodox among philosophers. And they have also been influential among mathematicians.[11] But we should note that the cognate suggestion by Russell that Euclid's arguments do not justify is—if we take it strictly and literally—drastically revisionary in principle of a traditional understanding of Euclid's geometry and its historical value. On a traditional view, the value derived from studying Euclid does not lie simply in detecting where he goes wrong. The implication of Russell's claim is, however, that we cannot come to know basic geometrical truths by studying Euclid. The use of the *Elements* as a textbook in schools over the centuries may have been justified by expediency, but it was not by itself a way by which students could genuinely acquire knowledge of geometry.[12]

We should also note that Russell's reading of Kant is quite problematic. To be sure, Kant was enormously impressed by the methodology and success of Euclid's geometry, and appears to be (perhaps culpably) ignorant of the longstanding doubts and debates about it. Otherwise, the strategy of the Aesthetic in this regard seems inexplicable. Moreover, Kant does not appear to have been an especially deep, and certainly not a creative, logician.[13] But he is no enemy of rigor: the point of taking Euclid as a model is in part precisely because his arguments are, as Kant sees it, exemplary of rigor.[14] Kant's appeal to intuition is not intended to be at odds with the need for rigor in mathematics; on the contrary, it is supposed to be part of the explanation of that rigor.

---

[11] See, e.g. Forder 1927, p. 42.

[12] A conclusion Russell himself drew (Russell 1901, p. 94-5). But we should also note that Russell was himself reacting against the sterility and insularity of the 19th Century Cambridge mathematics curriculum. In particular, there had long been heavy emphasis on rote learning of Euclid. See Rouse Ball 1905, p. 199ff.

[13] Though we can make too much of Kant's supposed neglect and complacency as to logic. It is sometimes suggested that Kant believed that logic did not need, and was not capable of, further development. But note that, at least as regards the first Critique, this is not what Kant actually says. He makes the point that logic (i.e. the logic of the syllogism) had not between his time and that of Aristotle been materially discredited or overturned, and then remarks that 'until now it has not been able to take a single step forward, and therefore *seems to all appearance* to be finished and complete' (Bviii; emphasis added). Of course, one might think of the logic of the syllogism as being discredited, in modern eyes, by its expressive limitations. But Kant's emphasis is, I suggest, on internal weaknesses or inconsistency. If so, then given that the logic of the syllogism remains little changed to this day, both the claims above were—and indeed are—surely true.

[14] Kant's respect for rigor is underscored by, for example, the terms in which he praises Christian Wolff in the B Preface, Bxxxvi.

From this perspective, Kant's discussion of one of Euclid's arguments in the Doctrine of Method is of great (and perhaps underappreciated) interest, for it represents a case study in which the reader is invited to consider, analyze and even test the respective contributions made by concepts and intuitions as Kant understands them. Moreover, the focus here is not so much on mathematical knowledge as such, as on mathematical reasoning. He is asking, in effect, the questions with which we started: how, if at all, can this kind of reasoning with diagrams justify belief? How, if at all, can it convey knowledge? As I shall describe below, Kant takes himself to be under a psychological constraint: to respect what the reasoner actually seems to be doing when she follows such an argument. And it is far from clear that predicate logic, even if it were available, would suffice to answer our questions in *this* context.

But this then raises a further worry. For it is part of the dialectic of the story told above—and reinforced by the logicist dismissal of intuition—that there are only two alternatives on offer here. *Either* the diagram is epistemically valuable, and we must postulate the existence of some special faculty of intuition; *or* the diagram is merely heuristic, and only non-diagrammatic or sentential arguments—and specifically arguments in a logical language—confer justification. Must we accept this choice, or believe that the reasoning here is empirical?

1.4 The Unacknowledged Alternative

I shall argue that the answer is No, and that the proffered choice in fact presents a false antithesis. There is a further and so far unacknowledged alternative: that the diagram can be epistemically valuable *a priori*, and yet require no special faculty of intuition to be so understood. This is the alternative I shall explore. Specifically, I shall defend the following three claims:

1. The kind of visual thinking we do in following an argument in Euclid can be epistemically valuable—and specifically, that it can justify belief and yield knowledge;
2. We can identify in the *Critique of Pure Reason* an embryonic account of such thinking that is preferable to its major alternatives; and
3. This account can be developed into a persuasive explanation of the epistemic value of this type of reasoning; one which is recognizably Kantian, but which does not appeal to any special faculty of intuition.

I briefly outline the course of the main argument below. But it may be helpful initially to note some of what is *not* claimed here. I do not claim that Euclid's arguments are proofs. Nor do I seek to defend Kant's principal claims in regard to Euclid from the Transcendental Aesthetic: that Euclid's geometry is, necessarily, the science of space, or that we can have synthetic *a priori* knowledge of physical space. Indeed I hardly discuss Euclid in relation to physical space at all.

Nonetheless, some general objections can already be anticipated to the very possibility of arguing in this way, at three progressively more inclusive levels:

- A first worry concerns whether and how diagrams can properly be used in the presentation of arguments at all, and whether they are in fact indispensable to such presentations; whether they cannot be dropped from these presentations without epistemic loss.
- A second worry accepts that diagrams can be used in the presentation of arguments, but is concerned with diagrams as spatio-temporal objects. On this view, reasoning with diagrams is and can only be a matter of gathering empirical evidence, and so Claim 1 above is committed to an empiricist epistemology. But this is allegedly implausible on other grounds as an account of mathematical reasoning.
- A third worry accepts that diagrams can be used in the presentation of *a priori* arguments, but claims that reasoning with diagrams in this way is not valid. In particular, diagrams do not contain instructions as to how they are to be understood. So reasoning with diagrams is unreliable, at least with respect to generalization.

We might also note a fourth line of objection, which attacks not the epistemic value of diagrams as such, but the claim that there is anything of distinctive epistemological interest here. It accepts that diagrams can be used in the presentation of *a priori* arguments, and that reasoning with diagrams in the relevant way is valid, but claims that it is so just in virtue of being logical reasoning. That is, though the diagram may be indispensable to a given argument, the justification conferred by that argument actually derives from the existence of a sentential proof of the same conclusion. It is not contested that logical reasoning can confer justification. But on this view, there is nothing epistemologically distinctive going on here.

Each of these objections, and others, is addressed in the discussion below.

1.5 The Argument of the Book

The goal of this book is, then, to explore whether and how the unacknowledged alternative can be positively elaborated, and defended against these objections. In order to do this, the discussion breaks down into three parts. The first part sets up the problem; the second explores candidate solutions; the third and longest part selects a preferred solution, refines it, defends it against a range of objections, and then develops it in what I take to be a plausible way. Specifically:

(I) Chapter 2 starts with a given argument as presented by Euclid and describes, in what I hope is an open and neutral manner, what a reasoner seems to be doing when she follows this argument in a certain way. To make the analysis and subsequent discussion as specific as possible, the discussion is focused on a single argument: Prop. I.32 of the *Elements*,[15] to the effect that all triangles have internal angles that sum to two right angles: the so-called 'angle sum' property. Chapter 3 sets out a logically exhaustive Framework of Alternatives, covering different theories that can be advanced to account for the apparent justification offered by this reasoning. This sorts such theories into four categories by their responses to two questions: if there is justification here, is it *a priori*? Does the diagram contribute, in a defined sense, to the justification?

(II) Chapters 4-6 describe and appraise three candidate theories that might be advanced in each of the categories identified by the Framework of Alternatives. These theories can be plausibly attributed to an interpretation of Plato by W.D. Ross, to J.S. Mill, and to Leibniz. Each theory holds that Euclid's argument confers justification, but they differ as to how it does so. Each chapter analyses the strengths and weaknesses of a given theory.

(III) Chapter 7 considers a fourth candidate theory, attributed to Kant; it argues that this account is superior to its alternatives, and defends it against what has come to be known as the Generality Objection. Kant's account has, however, been dismissed by even sympathetic interpreters as obviously mistaken. Chapter 8 isolates and defends a somewhat different but still recognizably neo-Kantian view against three main lines of criticism, in a way that highlights some of its distinctive features and commitments, as well as areas where

---

[15] Euclid's arguments are known as Propositions, and broadly follow a standard form described in Chapter 2 below. This term of art should not be confused with the meaning(s) of the same word in philosophy.

it diverges from the Kantian account already considered. However, a satisfactory explanation should not merely be superior to its alternatives. It should also satisfy further positive demands of logic and epistemology if it is to make good on the three claims above. Chapters 9 and 10 discuss these issues, and Chapter 11 concludes.

1.6 A Case Study

This book takes the form, then, of an extended case study of a single argument: Prop. I.32 of the *Elements*. The argument is well-known; indeed, even among Euclid's Propositions, Prop. I.32 had canonical status not merely with specialists debating the technicalities of geometrical reasoning in what became known as the *Quaestio de Certitudine Mathematicarum* in the 16th-18th Centuries, but also throughout more generalist philosophical debates as to the status of mathematics of the same period.[16] Virtually every major philosopher of the time discusses Euclid's geometry—including Descartes, Gassendi, Leibniz, Hobbes, Hume, Locke, Berkeley, and Kant—and most of them discuss Prop. I.32 in particular. By contrast, the general topic of diagrammatic reasoning has not been much explored in the recent philosophical literature. There has been an increasing body of work on this and related topics in cognitive and educational psychology, in the study of computing and artificial intelligence, and even in logic. However, there has been relatively little work on diagrammatic reasoning in mathematics, and no book-length treatment of the epistemology of diagrams as such.[17]

Proposition I.32 is perhaps too well-known for any further treatment to be really fresh and persuasive. But it is, I suggest, nevertheless a worthwhile choice, for three reasons. First, it employs an overall form of argument that is ubiquitous in Euclid's geometry, in which a diagram is constructed that represents a given situation, as to which a succession of logically interrelated claims are then made. So we have reason to think that conclusions reached here should generalize to other Propositions that use this form of argument, and perhaps elsewhere. Secondly, Euclid does not

---

[16] For details of the *Quaestio*, see Mancosu 1996, Ch. 1.

[17] Overall, see e.g. the collections Glasgow et al. 1995, and Blackwell 2001. Greaves 2002 gives a broad philosophical survey of diagrams in geometry and logic, but does not devote detailed consideration to the epistemology of reasoning with diagrams as such. For diagrams in computing/AI, see e.g., Sowa 1999 and Jamnik 2001; in logic, see e.g. the works of Barwise and his collaborators Etchemendy and Allwein, and Barwise's students Shin, Shimojima and Hammer listed in the References; and, for a case study comparing inference using diagrams and sentences in propositional logic, see Norman 1999. For diagrams in mathematics, see the works of Giaquinto, Manders and Brown.

here employ any distinctively questionable techniques such as superposition[18] or *reductio*, which might raise further questions for the present discussion before the basic approach has been assessed. On the view taken here, it would be a further step to argue that the diagram in a *reductio*, for example—in which there is no consistent set of claims to be represented by means of a diagram even in principle—can be epistemically valuable. This is a step I do not take, but the present discussion prepares the ground for it even so.[19]

Thirdly, we can take value from the historical pre-eminence of Prop. I.32 itself. In a different context Alasdair MacIntyre complains of

> the persistently unhistorical treatment of moral philosophy by contemporary philosophers... We all too often still treat the moral philosophers of the past as contributors to a single debate with a relatively unvarying subject matter, treating Plato and Hume and Mill as contemporaries both of themselves and of each other... Kant ceases to be part of the history of Prussia, Hume is no longer a Scotsman.[20]

Focusing on a single argument allows us, I suggest, to avoid this pitfall. Prop. I.32 has been preserved in more or less the same form since antiquity, and has been actively discussed in the modern era since the 17th Century. We can with confidence say that various different views expressed over time as to Euclid's arguments—and this argument in particular—address the same subject matter. This is not to say that the terms, leading concepts or background assumptions of participants have remained unchanged through the various debates in which Prop. I.32 has featured, of course. But it seems we have as good a case here as one could reasonably expect to test the value of this kind of comparison.

---

[18] A form of argument from the coincidence of lines or angles to their equality, as in Prop. I.4; widely regarded as philosophically suspect, even in antiquity. See Heath 1956, p. 225 ff.

[19] On reasoning via *reductio* in Euclid, see the helpful discussion in Manders 1995.

[20] MacIntyre 1981, p. 11.

2

The Euclidean Presentation

2.1 Introduction

This chapter examines Euclid's Proposition I.32, and the kind(s) of reasoning required to follow that argument. To follow this argument, a reasoner must be able to do at least two things: first, she must be able to understand the claims made at each stage, given appropriate interpretative conventions; and second, she must be sensitive to the validity of the transitions from premisses to conclusions.[21] This chapter discusses both these aspects of reasoning.

The chapter is divided into three parts. In part one (Sections 2.1-2.5), I introduce some relevant terminology, and briefly describe the general way in which arguments in Euclid are presented. I then set out Prop. I.32, and discuss the phenomenology, or introspective feel, of a given reasoner's experience as she works through the argument. In part two (Section 2.6), I examine the relevant conventions and assumptions lying in the background of Euclid's argument. In part three (Sections 2.7-2.9), I set out a detailed analysis of the argument, and discuss the role of the diagram in information representation and inference.

In this kind of analysis, it would be a mistake to assume from the outset that there is only one kind of reasoning to be explored. We need to leave open the possibility that two or more different types of reasoning can still constitute following a given presentation of an argument. We can then ask the question which, if any, of the inference types involved is compatible

[21] The notion of validity for inferences involving diagrams is discussed further in Chapter 10.

with the relevant phenomenology. However, the goal of this chapter is not just to situate Euclid's argument and the type(s) of reasoning involved in following it. It is also to bring out, in quite specific terms, the key questions with which subsequent chapters will be more generally concerned. For in order to assess various theories that purport to explain this reasoning, we need to be clearer as to what is to be explained. On only some of the possible candidate explanations will it be the case—even in principle—that the diagram has epistemic value.

As an introductory matter, it will be helpful to establish some relevant terminology. Following Harman, we can draw a distinction between implication and inference. *Implication* is a logical relation between, canonically, propositions, while *inference* is a type of mental act, whose outcome is a possible change in belief.[22] Specifically, I will take it that *an inference* is a transition between two (personal-level) mental states. More generally, I take *reasoning* to be a personal-level psychological process, consisting of inferences. In geometrical reasoning, we are reasoning about *geometrical objects* of a defined type (squares, triangles etc.).[23] These geometrical objects can have certain kinds of *property*, including shape properties (e.g. being right-angled). Such properties are described mathematically in Euclid in idealized terms that refer, for example, to perfectly straight lines without breadth. Geometrical objects can be represented by diagrams or figures. However, I will reserve the term *diagram* for physical inscriptions; when we visualize a geometrical object in imagination, I will call this a *figure*.[24] The visual properties of a diagram may represent geometrical properties, but they are not themselves geometrical properties, strictly speaking; a line on a diagram will never be perfectly straight, for example. This does not rule out a reasoner's judging of a diagram that 'that is square', but this judgment will be an observational (or, as it is sometimes more specifically called, perceptual-demonstrative) judgment, not a geometrical one; for the diagram will not be, as defined here, geometrically square.

Two further points. First, though some people refer to Euclid's arguments as proofs, it has been widely doubted whether they are in fact proofs; I shall simply refer to them as arguments, so leaving the further claim open.

---

22 Harman 1986, Harman 1999 Ch. 1.

23 The word 'object' here is just intended to denote the target of representation, and so should not be taken here in any metaphysically loaded way. Sometimes I shall refer generally to what is represented by a diagram as a 'situation'.

24 This has the advantage of avoiding a possible ambiguity in relation to the word 'figure', as between a diagram and an object or objects represented by a diagram. Note that, so as not to weary the reader by repeated reference to diagrams and figures, I will normally restrict discussion to diagrams.

Secondly, the discussion of reasoning will be concerned solely with the kind(s) of reasoning involved in *following* Euclid's argument, not with that involved in creating an argument or discovering a geometrical truth using Euclid's methods.[25] It has sometimes been suggested that thinking with diagrams can have epistemic value in regard to these processes. But this is not our topic.

2.2 Presentation and Argument

In logic and mathematics, it is often convenient to distinguish between different presentations of a given argument or subject matter. In geometry, one highly influential presentation of Euclidean geometry was given in Hilbert's book *Foundations of Geometry*.[26] But the most historically influential presentation of Euclidean geometry has, not surprisingly, been that of the *Elements* of Euclid himself, as we have it in Heiberg's text of 1883-8. This is a different presentation from that of Hilbert, in many respects: for example, it uses fewer and different axioms, and it has a quite different style of argument-presentation. One of the most striking differences is this: that presentations of arguments in the *Elements* use constructions of geometrical diagrams, and presentations of arguments in Hilbert do not.

We can, then, distinguish between *Euclidean* geometry (EG) and *Euclid's* geometry. I shall refer to the latter as the 'Euclidean Presentation' of Euclidean geometry, and where there is no risk of ambiguity I shall refer to Prop. I.32 itself as 'the Euclidean Presentation' (EP) or as 'Euclid's argument', for the sake of variation.[27]

The importance of the 'argument/presentation of argument' distinction lies in this, that when a reasoner is invited to follow a given argument, it is always a presentation of the argument that she follows.[28] But, unless the

---

[25] 'The reasoning involved in following' is a cumbersome locution, but it is preferable to short-cuts such as reasoning 'about' the relevant argument (which need not involve following the argument at all) or dubiously grammatical alternatives such as reasoning 'through' or 'with' the argument.

[26] Hilbert 1899.

[27] And note that the *Elements* does not in fact include all of what might now be considered Euclid's geometry. For example, the very striking '9-point circle' claim (that in any triangle, the midpoints of the three sides, the feet of the three altitudes, and the midpoints of the segments joining the three vertices to the orthocentre, all lie in a circle) does not appear in the *Elements*, and was only proven by Brianchon and Poncelet in 1820. Yet it can be derived using only the pure geometric methods of Books I-IV (see, e.g., Hartshorne 1997). There are many other results that are commonly considered results in Euclid's geometry, but that do not in fact occur in the *Elements*.

[28] This is not to say that such a presentation must always take the form of a physical inscription, like an argument presented in a textbook; it may be accessed via memory, for example.

individuation of arguments is abnormally strict, presentations of a given argument can significantly differ from each other, and can differ in the types of reasoning required to follow them correctly. So, if our target is an understanding of the type(s) of reasoning involved in following one particular presentation of a given argument, we cannot assume that this type of reasoning is the same as that involved in following a different presentation of the same argument.

2.3 The Euclidean Presentation

One could mean a variety of slightly different things by the term 'Euclidean Presentation', and there has been much debate over what is properly part of the *Elements*. Rather than engage in the relevant historical and exegetical questions, for the sake of convenience I will simply treat the standard English language text, Heath 1956, as canonical. We need only consider Book I. Later books include further definitions in Euclid's geometry, and indeed Euclid's arithmetic, but these are irrelevant for present purposes.

On this view, then, the Euclidean Presentation consists of, on the one hand, an initial set of definitions, common notions and postulates; and on the other, a set of 48 numbered Propositions.[29] Each Proposition contains an argument to one or more conclusions. Euclid uses several different broad forms of argument, including argument by superposition, such as in Prop. I.4, which argues from the coincidence of two triangles to one case of the side-angle-side claim for triangles;[30] argument by *reductio*; and argument by exhaustion. An example of the latter two can be found in Prop. I.19, which argues that in any triangle the greater side is subtended by the greater angle, from exhaustion of the alternatives; and as this brings out, a given Proposition can employ several forms of argument. The most common form of argument is by construction, however. Arguments by construction first give instructions for constructing a given diagram, and then make claims about the situation represented by the diagram.[31] Prop. I.32 contains an argument by construction, and it is on this form of argument that I will focus.

Regardless of the form of argument employed, Propositions in Euclid have a standardised and rather formulaic structure, which requires a brief

---

[29] Recall (from Section 1.6, fn.) that 'Proposition' here is a term of art in discussions of Euclid.

[30] Heath 1956, 306.

[31] Note that in some of Euclid's Propositions there are different cases to be considered, and each of these normally requires its own diagram. But Prop. I.32 is not one of these Propositions, and I will not discuss cases or case-branching here.

discussion. There are, normally, six divisions of a Proposition. Proclus describes these briefly as follows:

> Every problem and every theorem that is furnished with all its parts should contain the following elements: an enunciation (*protasis*), a setting-out (*ekthesis*), a specification (*diorismos*), a construction (*kataskeue*), a demonstration (*apodeixis*), a conclusion (*sumperasma*). Of these the enunciation states what is given and what is being sought from it ... the setting-out marks off what is given, by itself, and adapts it beforehand for use in the investigation. The specification states separately the thing that is sought, and makes clear precisely what it is. The construction adds what is lacking in the given for finding what is sought. The demonstration draws the proposed inference by reasoning scientifically from the propositions that have been admitted. The conclusion reverts to the enunciation, confirming what has been demonstrated.[32]

The notion of construction in Euclid has a relatively specific meaning. *Construction* is a process consisting of the application of Postulates 1-3. Postulates 1-3 contain instructions to the effect that, respectively, a straight line may be drawn from any point to any point; a finite straight line may be continuously produced (i.e. extended) in a straight line; and a circle with any centre and distance may be drawn (about a given point). Given a marker, these three operations can be executed on a flat surface by use of a straightedge and (collapsible) compass. *A construction procedure* is any finite sequence of these instructions, in any order. *A constructed diagram* is any diagram that results from execution of a construction procedure.

It is, in effect, a necessary and sufficient condition on a diagram in the Euclidean Presentation that it be constructible. Several important types of diagram are given separate definitions, such as triangles and squares, but they can also all be characterized in terms of different finite orderings of construction procedures, and Euclid does not generally use diagrams without giving previous constructions of them, or without its being obvious how they may be constructed.[33] In mentioning a specific type of diagram in the Setting-out of a Proposition, then, Euclid is using a kind of shorthand: no new information is being provided, over and above the construction procedure for that diagram.[34, 35]

---

[32] Proclus, p. 203; I have slightly amended the translation.

[33] Note that the same figure may in many cases properly be constructed using the same procedures executed in a different order. E.g. for an equilateral triangle constructed according to Prop. I.1, the order in which the sides are constructed will be irrelevant.

[34] Proclus, p. 204 claims that construction is often irrelevant on the grounds that 'in most theorems there is no construction because the setting-out suffices without any addition for proving the required property from the data.' But since the setting-out functions in these cases

Thus the construction procedure for an argument in Euclid is specified by a set of instructions indicating which postulate is to be applied, and in what order. It should be evident that this sense of 'construction' is not the same as that in which we sometimes talk of constructing an argument, for in the latter usage the contrast is with *following* an argument, and following Euclid's arguments often involves the reasoner in constructing a diagram. So I will reserve the term 'construction' for the sense described above, and where necessary use 'creating' an argument for the other sense.

2.4 Proposition I.32

With this in mind, we can turn to the argument of Prop. I.32, reproduced below. The square brackets are Heath's annotations; I have, however, labeled the various components of the proposition, and numbered the steps of the demonstration, for future reference.

> *Enunciation: In any triangle, if one of the sides be produced, the exterior angle is equal to the two interior and opposite angles, and the three interior angles of the triangle are equal to two right angles.*
>
> Setting-out (Construction): Let ABC be a triangle, and let one side of it BC be produced to D;
>
> Specification: I say that the exterior angle ACD is equal to the two interior and opposite angles BAC, ABC, and the three interior angles of the triangle ABC, ACB,[36] BAC are equal to two right angles.
>
> Construction: For let CE be drawn through the point C parallel to the straight line AB. [I.31]

---

by mentioning a figure, and *a fortiori* a construction, these cases utilize constructions, only without naming them as such.

[35] The definitions are, thus, verbal or 'nominal'; they do not go to the question of the existence or no of the object, which is apparently determined in Euclid by its constructibility. In determining what is to count as a construction, it seems the postulates are supposed to underwrite the existence assumptions of Euclid's geometry. But this is not necessarily to commit Euclid to what would now be considered a constructivist philosophy of mathematics. See Heath 1956, p. 143ff, and Mueller 1981, p. 15.

[36] Note that Heath has BCA here and below. BCA picks out the same angle as ACB, however, and ACB also appears in Heath's version. This variance is common in Euclid. However, since it is not clear what the purpose if any of this variance is, and since it may be confusing at first sight, I have used ACB for BCA throughout. Similarly, I have rendered CAB as BAC throughout.

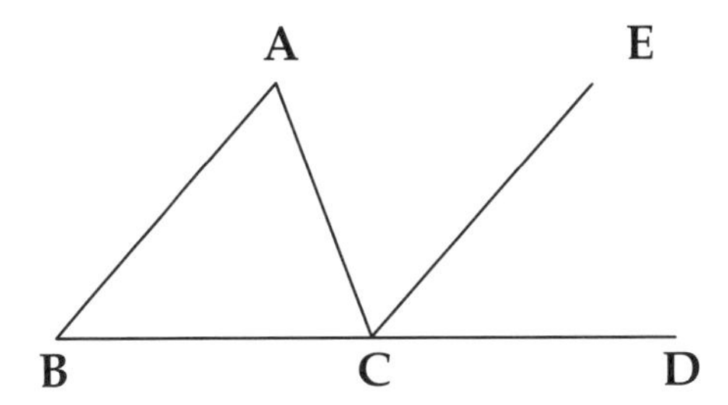

Demonstration:

I. Then, since AB is parallel to CE, and AC has fallen upon them, the alternate angles BAC, ACE are equal to one another. [I.29]

II. Again, since AB is parallel to CE, and the straight line BD has fallen upon them, the exterior angle ECD is equal to the interior and opposite angle ABC. [I.29]

III. But the angle ACE was also proved equal to the angle BAC; therefore the whole angle ACD is equal to the two interior and opposite angles BAC, ABC.

IV. Let the angle ACB be added to each; therefore, the angles ACD, ACB are equal to the three angles ABC, ACB, BAC.

V. But the angles ACD, ACB are equal to two right angles; [I.13] therefore the angles ABC, ACB, BAC are also equal to two right angles.

Conclusion: Therefore the exterior angle ACD is equal to the two interior and opposite angles CAB, ABC, and the three interior angles of the triangle ABC, ACB, CAB are equal to two right angles. QED.

2.5 Phenomenology

Let us take as standard the case in which a competent reasoner follows Euclid's argument carefully. In the course of this she does several different things: she is asked to draw a diagram, according to certain instructions given in the text or rubric; she reads sentences in the Demonstration, which express claims, some of which she is expected to understand, and assess the truth of, in relation to the situation represented by the diagram; and she is expected to be able to assess the validity of certain transitions in thought between these claims.

In the course of doing these things, the reasoner typically has a certain kind of experience. When one has a conscious experience, there is, at least normally, some felt quality to that experience. This is sometimes called the experience's *phenomenology*. Experiences A and B will be said to share the same (or similar) phenomenology when what it feels like for one to have A is the same as (or similar to) what it feels like for one to have B. On one conception of experience, certain perceptual experiences can have a phenomenology; and the same is true of conscious thoughts (i.e., occurrent, dated, episodes of thinking; including imaginative thinking). Importantly, we do not have to regard these thoughts as linguistically mediated (as having linguistic concepts as constituents, or as the result of operations using language).

The notion of phenomenology gives us a broad means to type different episodes of reasoning, at least on a prima facie basis. Two episodes of reasoning that have the same (or similar) phenomenology will be considered prima facie to be tokens of the same (or similar) type of reasoning. This typing is defeasible; there may be further grounds to differentiate between various types, and such grounds could be strong enough to cause us to doubt, and even to alter, an earlier classification. And typing by phenomenology is quite vague, though it may be hard to make more precise. But it supplies a useful pre-analytic means to identify different types of reasoning as such. As a result, considerations of phenomenology can impose a strong though defeasible constraint on possible explanations of the reasoning involved in following Euclid's argument. We will want to require of a candidate explanation that it be psychologically realistic; that is, faithful to the distinctive character of this reasoning, insofar as that can be assessed.

I suggested above that a competent reasoner who follows the Euclidean Presentation carefully has a certain typical kind of experience. To say this is to presuppose a rather broad notion of experience, not restricted to experience of the external world, on which conscious thinkings and imaginings can count as experiences. But such a broad notion is surely quite intelligible, even commonsensical.[37] Can we describe the phenomenology of the reasoner's experience here?[38] I have already mentioned two aspects: it includes a visual experience of (or as of) a diagram; and it includes the experience of taking certain claims made in the text to be true of what the diagram represents. But I suggest that we can also identify three other distinctive aspects.

---

[37] I will discuss this and narrower conceptions in more detail below, especially in Chapter 3.

[38] Again, I restrict attention to someone who follows the argument in relation to a diagram, rather than an imagined figure. But note that a further way to follow Euclid's argument can be via visual memory. That is, one can recall seeing, for example, the page of a textbook containing the argument, and then follow the argument by recalling reading the text and reasoning with the figure accordingly.

- First, a feeling of *accessibility*: the reasoner seems to understand each inference, and the argument as a whole, very readily. The argument is short, and the line of thought is evident. There is little or no jargon or appeal to specialist knowledge here. There is little felt need to scrutinize details of the diagram, which might obstruct the flow of thought: the reasoner does not measure the angles of the diagram, she does not measure the lengths or assess the straightness of the sides.[39]
- Second, a feeling of *certainty*: the reasoner seems to feel a strong conviction at the end in the truth of the general conclusion. This conviction might grow slightly with a little further reflection, but it does not grow thereafter. It is not diminished by any concomitant recognition that the diagram may not be perfectly drawn, or may only be roughly similar to a geometrical triangle.
- Third, a feeling of *clarity*: the reasoner has gained not merely belief or knowledge of a general truth, but a feeling of understanding: not merely that the conclusion must be so, but of why it must be so. By coming to understand the general relationship between the angles formed by the auxiliary line CE and the opposite angles of the triangle (∠ABC, ∠BAC), she grasps that the angles of any such triangle must equal the angles on a straight line, without exception.

The historical record suggests that this phenomenology should be taken as a datum. Many people have had this experience, and some of these features have been noted by other writers on geometrical thinking.[40] It is clear that, like other arguments in Euclid, Prop. I.32 has long been regarded as extremely accessible, convincing and clear by many reasoners.

Now the description above is clearly consistent with more than one way in which a reasoner might follow Euclid's argument. Consider someone relatively new to geometry who, familiar with the definitions, common notions and postulates of Euclid, approaches Prop. I.32. It seems that she might follow the argument thus: she could initially take the sentences in the demonstration to be about a triangle or class of triangles that the diagram

---

[39] Note that the feeling of accessibility I have in mind here is not the same as the feeling of immediacy that one can sometimes have on perceiving a diagram. For example, experience of a Gestalt 'switch' in some visual illusions using diagrams can give a feeling of sudden recognition or insight on the viewer's part, and it may be that this involves no mediating inference. However, distinguishing between accessibility and immediacy does not rule out that a reasoner following Euclid's argument can nevertheless experience a rush of understanding or insight as a result, or that certain types of inference may employ such 'switching'.

[40] E.g. Giaquinto 1992 (in relation to similar processes used in discovery by visualization), which has influenced this discussion; cf. Brown 1999, Ch. 3.

depicts, i.e., to which she takes it to be relevantly similar in visual appearance.[41] When she reaches the end of the argument, she could form the belief that the argument succeeds for the class of triangles depicted. If she then reflected that no step in the argument depended on any property of the triangles depicted that was not a property of all triangles, she could form the general belief that the argument succeeded for all triangles.

Call this the *naïve approach*. We can contrast it with the *expert approach*; for it seems that a more expert reasoner might take the diagram from the outset to relate to triangles generally (as determined by the text), including those triangles to which the diagram bears no visual resemblance. Unlike that of the novice, the expert's final generalizing inference is not reflective; the conclusion she reaches after following the argument in relation to the triangles represented by the diagram will already be a general one.

The novice's and the expert's reasoning have slightly different phenomenologies. For the expert, there is no feeling that the diagram depicts any particular type(s) of triangle, and there is no conscious experience of any final generalizing inference. For the novice, by contrast, the phenomenology of the experience may be this: that she is 'seeing what happens' to a diagram of a triangle or class of triangles, and tracking it through a sequence of changes, before a final generalizing inference.

For the sake of specificity I will focus on the expert approach; and it will be convenient to refer to this in places below as 'the' reasoning involved in following Euclid's argument. But, to reiterate, this should *not* be taken to imply that there is only one way to follow Euclid's argument. Quite the contrary; I suggest that the naïve approach is also available, and indeed—as I discuss in Chapter 9—there may also be other approaches. The contrast between the naïve and the expert approaches is a useful one, however, and it will be further developed below as part of an attempt to explain the nature of the cognitive achievement that constitutes mastery of Euclid's argument.

Following Euclid's argument, however, requires a reasoner to grasp various tacit or explicit background assumptions and conventions, in particular as relates to the construction of, and claims made as to, the diagram. To these I now turn.

---

[41] What this might amount to is discussed further below, and in Chapter 9.

2.6 I.32: Background Conventions and Assumptions[42]

The construction of the diagram in Prop. I.32 takes place in two stages, as we have seen: in the Setting-out, and in the Construction.[43] To carry out the construction properly, the reasoner must be able to draw a triangle and label it; to extend line BC to D, and label D; and to draw auxiliary line CE and label E. That is, she must understand (1) how the diagram is to be drawn, and (2) how it is labeled. If she is then to follow the demonstration successfully in relation to the constructed diagram, she must understand (3) the representational scope of the diagram, and (4) any other relevant properties or features of the object(s) represented. All four aspects are controlled by various background conventions and assumptions, and I will discuss these in turn.

Drawing the Diagram

Prop. I.32 leaves unspecified exactly how the diagram is to be drawn: the initial instruction is simply 'Let ABC be a triangle.' This can be done in various different ways in Euclid. At its simplest, a diagram of a scalene triangle may be constructed by drawing a line segment BC (according to Postulate 1), drawing circles that cross at some point A not on BC or its extensions (using Postulate 3), then drawing further lines AB and AC (using Postulate 1 again). Alternatively, it may be equilateral (constructed as per Prop. I.1; I give the procedure below) or isosceles (a simple variant construction). Having constructed the initial diagram, the reasoner can then amend it as instructed. To produce line BC to D, she must be aware of Postulate 2; and to draw the parallel line CE, of the construction procedure in Prop. I.31.

Constructing the diagram, then, relies on a range of background assumptions: for example, that the reasoner has the relevant concepts of 'point', 'line' and of 'straight line'; that she understands the Postulates; that the straightedge and compass are used as instructed (e.g. that the reasoner can insert the compass into each end-point of BC); that the relevant points B and C exist between which line BC can be drawn.

It might seem as though awareness of these assumptions would be enough to allow the reasoner to construct the diagram appropriately. In

---

[42] I owe several of the leading ideas in this section to conversations with and work by Marcus Giaquinto.

[43] Euclid brings part of the construction process into the Setting-out because the construction of exterior angle ACD is required to make his claim in the Specification about it intelligible. CE is an *auxiliary* line because, unlike the extended portion of BCD, it is not required to make the claim to be argued intelligible, and it does not appear in that claim, and so not in the conclusion.

fact, however, this is not the case, as the reader can check. For the following three incorrect diagrams are all permissibly constructible from the instructions given in Prop. I.32, supplemented by these assumptions:

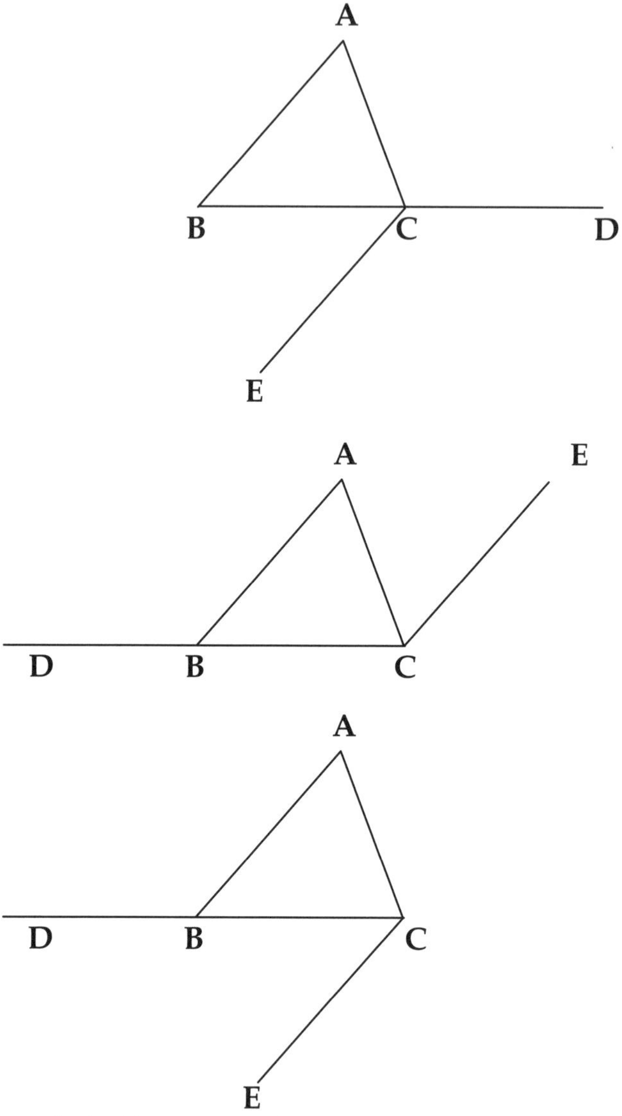

Why is this so? There is a further implicit assumption behind the construction procedure here (and elsewhere in Euclid): it is not stated in which direction BC is to be extended to D, nor CE drawn relative to BC. These together create four alternatives, of which only one is intended. Euclid's argument cannot proceed in relation to the other diagrams, but he does not explicitly supply the relevant conventions required to rule them out.

In Heath's edition, the relevant information is provided by showing the correct intended diagram. Strictly speaking, however, further explicit conventions are required, to rule the deviant diagrams out. The latter two diagrams can be ruled out by noting the implicit convention in Euclid that the ordering of letters in lines determines their direction: these would be instances of line CB, not line BC, being produced to D. What about the first diagram? We might rule this out in one of two ways: either by defining a general convention governing the drawing of parallel lines, or by giving a specific instruction here—not supplied by Euclid—in the text of the construction procedure. The first route looks unattractive; why should it not be convenient to draw CE down as in the first diagram, for other arguments? So what we need here is a specific instruction: that E is to lie on the same side of BCD as A.

What this brings out is that the modern tendency to start with a drawn diagram, and consider only that plus the demonstration as Euclid's argument proper, is just a short-cut to an outcome that should, strictly speaking, be reached via the appropriate construction procedure, given background conventions including that (or similar) above.

Labeling the Diagram

Secondly, there are background labeling conventions and assumptions. We do not know whether Euclid in fact labeled the diagrams in the *Elements*, though some ancient manuscripts have labels; but some method is needed to identify and track vertices, lines and angles in the diagrams constructed, and labels serve this purpose, as in Heath 1956. Here are some of the labeling conventions at work in Prop. I.32. A large Roman letter 'X', 'Y', 'Z'... next to a vertex is a label for that vertex. A letter repeated in the text refers to the vertex labeled by that letter; the letters 'XY' in the text refer to the line lying between the vertices of the diagram labeled 'X' and 'Y'; 'angle XYZ' refers to the angle between the lines labeled 'XY' and 'YZ' at the vertex labeled 'Y', etc. Repeated labels always refer to the same vertex, line or angle.

Representational and Depictive Scope

The above conventions govern the drawing and labeling of the diagram, and the relation between the diagram labeled and the text of Euclid's argument. However, a drawn and labeled diagram is merely a collection of marks, of letters and lines. What makes the diagram a diagram *of something* is its representational content, and this is constrained by a further set of representational conventions and assumptions.

We need to distinguish here between depictive scope and representational scope. We can think of *depictive scope* as governed by the following convention:

(C1) The reasoner is to take a diagram to depict all those geometrical objects that it appears to her visually to resemble.

Thus drawn lines are to be taken to depict geometric lines, points to depict geometrical points, shapes to depict geometrical shapes. Note that a further convention can be given in the other direction, in relation to construction:

(C1*) The reasoner is to take an instruction to construct a geometrical object of a certain kind as an instruction to construct a (saliently sized and clearly drawn) diagram that appears to her visually to resemble an instance of that kind.

On the other hand, we can think of *representational scope* as governed by the following convention:

(C2) The reasoner is to take a diagram to represent all those geometrical objects that can in principle be generated by execution of the construction procedure specified in the text.

It is an error to take the diagram to represent only those objects that fall within its depictive scope at the conclusion of Euclid's argument: the conclusion is supposed to be true of all triangles, not merely of those that the diagram visually resembles. But note that this does not prevent a reasoner from using (C1*) to draw the diagram. There is nothing in (C1*) that requires a reasoner to treat a given diagram as having merely depictive scope.

As this brings out, it is the text of Euclid's argument—and specifically the relevant construction procedure—that controls the representational scope of the diagram, and the text may require the diagram to be understood as representing classes of geometrical objects that it may not appear to the reasoner to depict. A diagram can in principle be taken to represent anything. It does not carry with it its own instructions; rather, these must be supplied from elsewhere.

These conventions, and the controlling influence of the text of the argument, can be seen at work in Prop. I.32. The reasoner may, in constructing the diagram, construct a diagram of an equilateral triangle. However, the instructions in the argument ('Let ABC be a triangle') do not require that an equilateral triangle be drawn: she might in principle have obeyed them in constructing an isosceles or scalene triangle. Hence, for her to treat the diagram as restricted to equilateral triangles is to make an additional assumption, and one that has no warrant in the text.[44] Similarly, the color of the diagram is left open; it may be any color, and so which color it in fact is, is irrelevant. Similar remarks apply to the size of the diagram; it may in principle be any size, and so what size it in fact is, is irrelevant.[45] Here the modern tendency to omit the construction and treat the argument as beginning from an already completed figure renders Euclid's argument unintelligible, unless further and equivalent scope conventions are also supplied.

Distinguishing between representational scope and depictive scope allows us to unpack the difference between the naïve and expert approaches described above. The naïve reasoner takes the representational scope of the diagram to be determined solely by its depictive scope. But later, in reflecting that no step in the argument depends on the specific properties of the depicted triangles, she does not take the conclusion to be limited merely to those triangles that lie within the depictive scope of the diagram. It is then a further movement of thought for her to recognize that the representational scope of the diagram can be entirely general, as does the expert. Until then, it is natural, and not mistaken, to take the diagram to be depictive, and it is in this sense that she can be said to be 'seeing what happens' to it, and to the object(s) it depicts.

The Target of Representation

Finally, there are assumptions governing the geometrical objects represented by the diagram. It is appropriate to distinguish these assumptions from those relating to the representational content of the diagram, because some intended properties of the geometrical objects represented may not be detectable from a visual examination of the diagram, and yet affect the cogency of Euclid's argument. We can illustrate this by briefly examining a

[44] I discuss questions of construction in more detail below, especially in Chapters 7 and 10.

[45] It is sometimes objected that size is in fact relevant; even if space is (as perceived) locally Euclidean, since the physical universe appears to be non-Euclidean, very large diagrams will, if they follow the shape of space, violate the parallel postulate. But this worry is irrelevant here, for what is in question is how objects represented by the diagram would be if the relevant space were Euclidean. We can and do follow arguments in Euclid's geometry containing diagrams that are not themselves perfectly Euclidean, even when we know in advance that they are not.

well-known objection to the construction of an equilateral triangle in Prop. I.1, as below:

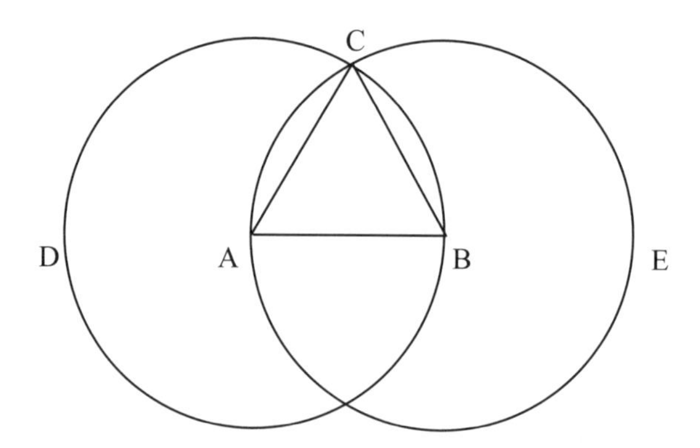

The construction procedure given in Prop. I.1 is as follows:

> Setting-Out [Construction]: Let AB be the given finite straight line.
>
> ...
>
> Construction: With centre A and distance AB let the circle BCD be described; again, with centre B and distance BA let the circle ACE be described; and from the point C, in which the circles cut one another, to the points A, B let the straight lines CA, CB be joined.

It is often pointed out that Euclid is not entitled to assume the existence of the intersection point C between the two circles D and E. It appears that Euclid has a notion of continuity, perhaps given in visual perception by the apparently unbroken motion of a stylus or pen that is taken to represent a mathematical point, on which it is obvious to him that such a point C exists. However, we can provide models of Euclid's axioms in discontinuous geometries in which point C does not exist; it has been argued that what is required to fill this gap is a further assumption, akin to a postulate, to the effect that the relevant concept of continuity is, or is a derivative of, Dedekind continuity.[46]

---

[46] A line is Dedekind-continuous if it satisfies the following condition (Dedekind 1963, p. 11): 'If all points of a straight line fall into two classes such that every point of the first class lies to the left of every point of the second class, then there exists one and only one point which produces this division of all the points into two classes, this severing of the straight line into two portions.' See also the discussion in Heath 1956, p. 234ff.

Now, there is something correct about this view: there is nothing as such about the visual features of a drawn line that instructs a reasoner to take it as representing a continuous (or non-continuous) geometrical line. A drawn line is, by itself, representationally neutral as between these alternatives. So what warrants the existence of C is not the visual appearance of the diagram, but a further assumption about continuity; and the soundness of the argument in Prop. I.1 relies on this. The example, then, correctly brings out the insufficiency of the visual features of a diagram alone, even when its representational scope is correctly grasped by a reasoner, to convey information as to all the properties of the object(s) represented.

Nonetheless, it would be a mistake, as well as an anachronism, to require of Euclid the assumption of Dedekind continuity. For a suitable continuity assumption for circles here can be simply stated: if circle D has one point inside and one point outside circle E, then the two circles intersect in two points. And a point P can be defined as inside a circle with centre A and radius AB if AP < AB (and outside if AP > AB).[47] This assumption does not require modern logic to be grasped by a reasoner. Euclid simply seems to take it for granted.

If we take Euclid's goal to be the project of presenting a fully explicit deductive geometry, then the absence of a suitable and explicit continuity assumption is not a trivial weakness. In a fully explicit presentation of Euclid's argument, this and other background assumptions and conventions, including others not mentioned here,[48] would be stated outright. Some of these are given already explicitly, in the Postulates, Common Notions and Definitions, though some of these themselves require supplementation and emendation, as has often been noted. Other assumptions are given implicitly and could be derived from a careful reading of Book I: for example, the convention that an unconstructed diagram is not to be used in inference is implicit in Euclid's own standard practice.[49] Still others must be supplied from outside Euclid; as, for example, with the continuity assumption discussed above. Carrying through this program in detail is beyond the scope of this discussion, and the complexity involved would itself generate other possible sources of error; but there is no reason in principle to think that it cannot be done.

---

[47] See the discussion in, for example, Greenberg 1993, p. 94.

[48] Meserve 1955 (p. 231) usefully lists some further tacit assumptions in Euclid.

[49] Hence, on some views, Euclid delays Prop. I.32, which uses the construction of line CE parallel to AB, until after Prop. I.31, which sets out this construction for the first time.

2.7 Analyzing the Argument of Prop. I.32

With this in mind, we can now use a reconstruction of Prop. I.32 to analyze each of the steps of the argument, and the interplay between the diagram and claims made about the situation that the diagram represents, in more detail. Here is such a reconstruction, in a more modern format.

Claims (Specification):

For any triangle ABC on line BCD
1. ∠ACD = ∠ABC + ∠BAC
2. ∠ABC + ∠BAC + ∠ACB = two right angles.

Setting-Out:

Let ABC be a triangle [by Definition 19]

Construction:

Let BC be produced to D [by Postulate 2]
Let CE be drawn through C parallel to AB, E to lie on the same side of BCD as A [by Prop. I.31][50]

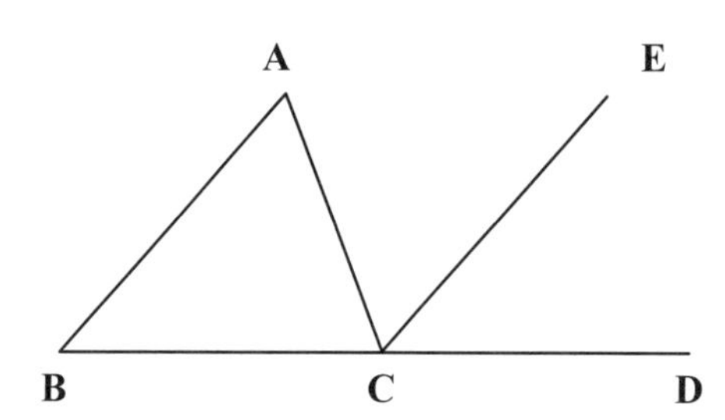

[50] Note that Prop. I.31, and not Postulate 5, shows that through a given point a straight line can be drawn parallel to a given straight line, which is what is required here.

Demonstration:

(1) ∠BAC and ∠ACE are alternate [from the diagram]
(2) Alternate angles are equal [by Prop. I.29]
(3) ∠BAC = ∠ACE [1, 2: by substitution]

(4) ∠ECD is exterior and opposite to interior ∠ABC [from the diagram]
(5) Exterior and interior opposite angles are equal [by Prop. I.29]
(6) ∠ABC = ∠ECD [4, 5: by substitution]

(7) ∠ACD = ∠ECD + ∠ACE [from the diagram]
(8) ∠ACD = ∠ECD + ∠BAC [3, 7: by substitution]
(9) ∠ACD = ∠ABC + ∠BAC [6, 8: by substitution]

Line 9 is Claim 1.

(10) ∠ACD + ∠ACB = ∠ABC + ∠BAC + ∠ACB [9: by CN2, adding ∠ACB to both sides][51]

(11) ∠ACD + ∠ACB is the sum of all the angles on BCD [from the diagram]
(12) BCD is a straight line [by the first Construction step]
(13) ∠ACD + ∠ACB is the sum of all the angles on a straight line [11, 12: by substitution]
(14) The sum of all the angles on a straight line = two right angles [by Prop I.13]
(15) ∠ACD + ∠ACB = two right angles [13, 14: by substitution]
(16) ∠ABC + ∠BAC + ∠ACB = two right angles [10, 15: by substitution]

Line 16 is Claim 2.

The 5 steps of the Demonstration in Euclid are indicated by the five clusters of sentences—1-3, 4-6, 7-9, 10 and 11-16—in the above reconstruction. So it should be evident that, although it differs in making some of the inferences more explicit than Euclid does, this reconstruction still presents Euclid's argument, or something very similar.

It is interesting to compare this argument with the Rouse Ball fallacy analyzed in Section 1.2. Here again, many of the lines of the argument are

[51] Recall that Common Notion 2 states that 'if equals be added to equals, the wholes are equal.'

warranted either by already-established results (e.g. line 2), or by established rules of logic (e.g. line 3, by substitution of co-referring terms), or by the text of the construction rubric (e.g. line 12, which is warranted by the first construction step).[52] So far, so good; it is uncontroversial that a reasoner is entitled to take these claims as known. But some of the lines of this argument—lines 1, 4, 7 and 11—are warranted by the diagram. This is more problematic. How can one's experience of a diagram justify a belief state about something other than the diagram?

Diagrammatic Information

The answer to the question above is this: a diagram may have representational content. For example, it may represent one circle as intersecting another, as in the previous section; or it may misrepresent, as the first three diagrams in Section 1.2 misrepresent the situation described in the rubric of the Rouse Ball fallacy. But note that this is not to claim that diagrams have (or cannot have) propositional contents. The thought being defended here is, not that diagrams have propositional contents as such, but that in certain contexts a proposition can be inferred or known by observing a diagram: that someone can reliably infer the truth of a given claim about a situation represented by a diagram from observing such a diagram.

Sometimes a diagram's contribution is not strictly propositional. Recall that the construction rubric says 'Let ABC be a triangle…'. We can compare a similar type of sentence to be found in many sentential arguments: 'Let X be a prime number', or 'Let's say Y is an average taxpayer.' These sentences do not present propositions, even when represented in the form 'Let it be the case: X is a prime number.' They do not express claims that could be correct or incorrect, and there is no correct response to them along the lines of 'No, X isn't a prime number', or 'It's true, Y *is* an average taxpayer.' Rather, such sentences serve to introduce a term denoting an entity of a given kind into the argument, as to which certain claims will subsequently be made. It seems that the same or a very similar function is being played by the 'let …' sentences of the construction rubric here; they are introducing a term for an object (or class of objects) of a particular kind represented by the diagram, as to which certain claims will later be made.[53]

If this is right, then we do not need to see the diagram as conveying propositional information when it is introduced. Once it has been introduced and labeled, however, propositional claims can be made about the

---

[52] Note that, unlike that in Chapter 1, this argument also uses Euclid's Common Notions directly (e.g. line 10).

[53] I discuss the relation between the diagram and the 'let…' premiss further in Chapter 9.

situation represented by the diagram, and such claims can serve as the premisses of inferences.

Knowledge from the Diagram

How, then, can a reasoner come to know line 1 of the Demonstration from the diagram? One way in which she might is this: she might observe that the diagram correctly represented the situation described in the rubric, and that line 1 was true of that situation. On this view, if she took the diagram to be a triangle with an extended base and an auxiliary line, if she could reliably identify the angles in question on the diagram and if the diagram so understood correctly represented the situation described by the rubric, then she might see from the diagram that the relevant angles in the situation represented were alternate. Given knowledge of enough of the relevant background conventions and assumptions, she can readily satisfy the antecedents in this conditional. For she has correctly constructed the diagram, following the rubric, in order to represent the situation described. So she can come to know line 1 by this means.

Similar remarks apply with respect to line 4. But what about line 7? Line 7 claims that ∠ACD = ∠ECD + ∠ACE; that is, it states an equality of size between one angle and the sum of two others. How can a reasoner come to know this? Here is one way: line 7 might be reached by a further inference as follows:

(7a) CE divides ∠ACD into two parts, ∠ECD and ∠ACE, without remainder [from the diagram]

(7b) The whole of an angle is equal in size to the sum of the sizes of any parts into which it is divided without remainder [background assumption]

(7c) ∠ACD = ∠ECD + ∠ACE [7a, 7b: by substitution]

The background assumption in line 7b here is independently plausible. A reasoner who has the relevant concept of addition can know line 7a by observation of the diagram given her background knowledge that, in the situation represented by the diagram, line CE is breadthless (by the definition of 'line'). Given these premisses, 7c = 7 follows straightforwardly.

Is this the only way in which she can come to know line 7? I suggest not. Say (P1) we recall Common Notion 4, that 'things which coincide with one another are equal to one another', and treat this as entitling a reasoner to conclude from the coincidence of two sets of lines that the angles between them are equal in size. Suppose further that (P2) we treat different ways of seeing a single diagram as equivalent to seeing two diagrams known to be exactly coincident with each other. Given suitably explicit background conventions, both these principles are not, I suggest, implausible.[54] Then it seems that a reasoner might come to know line 7 by looking at the diagram in two different ways: first, seeing the two lines AC and CD as containing ∠ACD alone, intersected by CE (which can be ignored, since its presence or absence does not affect the size of ∠ACD); and secondly seeing the same lines as containing ∠ECD and ∠ACE and nothing else. This situation can be represented in the subportion of the diagram given below, where the angles between the dotted and whole lines are the relevant angles.

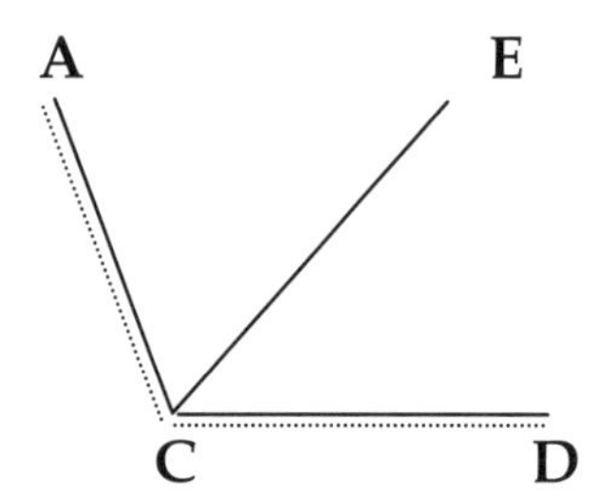

But then by (P2) above, seeing the diagram in these different ways is equivalent to seeing two exactly coincident diagrams, and by (P1) above, the reasoner is entitled to conclude that angles ∠ECD + ∠ACE and ∠ACD are equal in size, and so to believe line 7.

Note that during this process nothing has changed about the diagram. What seems to warrant the inference, rather, is an act of 'seeing as'. It is by seeing the diagram as presenting information as to the size of angle A, and then seeing the diagram as presenting information as to the size of angles B and C, together with (P1) and (P2) and background knowledge that the diagram has not altered, that the reasoner is entitled to come to believe that for any angles A, B, C: angle A is equal in size to angles B and C.[55]

---

[54] As noted above, CN4 was regarded as suspect even in antiquity, and Euclid appears to avoid it where he can. Historic worries have standardly related to the tacit introduction of assumptions as to rigid translation, to which no appeal is made here, even tacitly.

[55] On different kinds of 'seeing as', cf. Budd 1989, Ch. 4. The kind of 'seeing as' that we can detect here does not involve the strongly different and sometimes mutually exclusive experiences characteristic of seeing, e.g., a duck vs. a rabbit, or a Necker cube oriented in different ways. But there is nevertheless an experience here of 'flipping' between two different ways of seeing the lines in question.

We can give a similar analysis of line 11, the claim that ∠ACD and ∠ACB is the sum of all the angles on BCD. Again, one way to come to know this is by a substitutional inference, as follows:

(11a) AC divides the angle on line BCD into two parts, ∠ACD and ∠ACB, without remainder [from the diagram]

(11b) The whole of an angle is equal in size to the sum of the sizes of any parts into which it is divided without remainder [= 7b; background assumption]

(11c) ∠ACD + ∠ACB is the sum of the sizes of all the angles on BCD [11a, 11b: by substitution]

And again, there is an alternative visual means to come to know line 11 analogous to the process already described, which exploits the different ways in which a reasoner can observe the relevant portion of the diagram: seeing it either as a whole angle at C formed by line BCD on the same side as A, or seeing the same lines as containing two angles ∠ACD and ∠ACB.

We can, then, describe three ways in which a reasoner can use the diagram to come to know a claim in the argument. First, she can see that the claim is true in relation to the situation represented by the diagram (e.g., lines 1 and 4). Secondly, she can infer the claim from the conjunction of a claim taken from the diagram and a further (perhaps not visually known) claim, using a substitutional form of inference (lines 7c and 11c). Thirdly, however, she can infer the claim from different ways of seeing the diagram, given a suitable inference rule such as (P2) for 'seeing as' (lines 7 and 11, reconstructed above). Note that this last inference is not substitutional, is not general in nature but apparently limited to geometry, and seems to require the reasoner to reason with the diagram: it is by 'flipping' between different ways of seeing the diagram that a reasoner is enabled to make the inference.

Which of these latter two routes best fits the phenomenology described earlier? We need to note the important fact that Euclid's argument does not in fact mention either of the claims in lines 7 and 11 as such; it just passes over them. The reconstruction clearly indicates that they (or something similar) are required for the argument to go through, however, and a careful reasoner would detect that there is a lacuna between, say, the first and second halves of line (III), of which line 7 is an important component:

(III) But the angle ACE was also proved equal to the angle BAC; therefore the whole angle ACD is equal to the two interior and opposite angles BAC, ABC.

Just to speak from personal experience: when I read (III) in context, I seem to focus on the diagram. I have difficulty even understanding (III) without looking at a diagram, or visualizing a figure, and—unless I am talking through the argument or self-consciously framing my thoughts in language at each stage—I am not sure I consciously think thoughts with linguistic contents at all in relation to the intervening claims here. When I think through the transition required to reach line 7, I do not seem to do any substitutional reasoning; nor do I entertain any conscious thought corresponding to the major premiss in line 7b above. Indeed, I do not seem to use a conscious process of inference at all to reach line 7. Rather, I look at the diagram and derive the information in a way that is phenomenologically immediate, or almost so.

The third, visual, inferential route described above can in principle explain this feeling of accessibility, since it used a kind of 'seeing as', and this is often associated with swift, and even apparently immediate, inference. This is hardly conclusive, and of course other people may have phenomenologically different experiences. But the wider point is that there are several valid routes to belief here; and the visual route I have just described is one of them.

Visual Inferences

So far we have identified two distinct inference types, one that operates by making substitutions on sentences, and one that operates by seeing the diagram in different ways. Are there any others available here?

The reconstructed argument above suggested an underlying similarity of inference-structure in four of the five steps of the Demonstration in Prop. I.32. For in each of these cases the inference proceeds from two or more premisses, one of which is known in virtue of the diagram, to a conclusion, via substitutional reasoning. The exception is the inference to line 10, which has no premiss taken from the diagram.

However, the analysis above suggests that it is not compulsory to regard the inferences between lines of the argument as substitutional. If we accept that there is a visual route to knowing line 7, as described above, then—for example—we do not have to regard the inferences to lines 8 and 9 as inferences by substitution of terms in sentences. A substitutional inference is in both cases in principle available to the reasoner, since the equalities of angles are already given in lines 3 and 6. But it also seems that a reasoner could reason with the diagram to the required conclusion: to reach line 9 from line 8, for example, she might visually imagine a copy of ∠ECD sliding rigidly along line BCD until it was mapped it on to ∠ABC. Given that the angles involved are equal, and already known by the reasoner to be so, such a visual inference will not go wrong. Moreover, it is a different type of visual inference to that of 'seeing as' used to reach line 7; there is no seeing of the same set of lines in two different ways. And whereas the visual route to line 7 was, in effect, an inference to an equality of angles from their coincidence, this inference is one of visual translation of angles already known to be equal.

Similar remarks can be made about the inference to line 10, the claim that ∠ACD + ∠ACB = ∠ABC + ∠BAC + ∠ACB. This could properly be inferred, as indicated, by a form of addition on sentences applied to line 9, given background knowledge of Common Notion 2 ('If equals be added to equals, the wholes are equal'). But it could also be known by an inference using the diagram. Take someone who reasons from the diagram to line 9, that ∠ACD is equal to the sum of the sizes of the opposite internal angles of triangle ABC, as below.

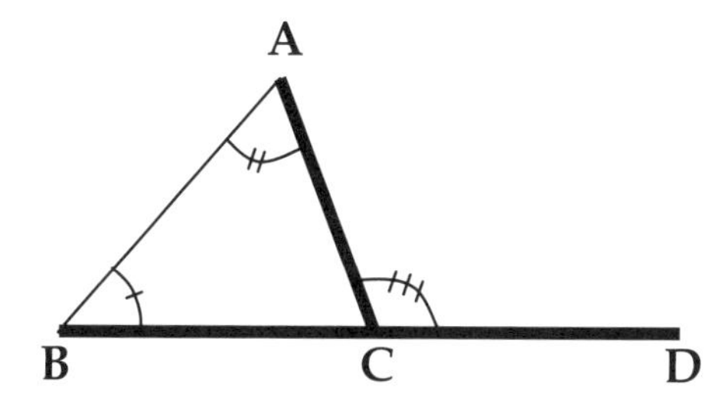

The reasoner might then observe from the diagram that ∠ACB is common both to triangle ABC and the angles on line BCD (given that AC is breadthless), as marked by the boldface lines. But she can then visually imagine adding ∠ACB to both, to reach the desired result that ∠ACD + ∠ACB = ∠ABC + ∠BAC + ∠ACB. Again, this is a distinct form of visual inference: it is not an inference of 'seeing as', and though it is additive, it does not involve visual translation of angles.

So far, then, I have contrasted substitutional inference on sentences with three distinct types of visual inference using diagrams: using 'seeing as', using visual translation, and using visual addition. Note that *both* broad categories of inference—substitutional and visual—seem to require a diagram, at least to supply a reference for the various labels in the claims or sentences entertained in thought, so that the reasoner can detect what the argument is about. This was brought out by the difficulty of even understanding line (III) above from Prop. I.32 without reference to a diagram.

But there is a key difference between the visual and substitutional categories of inference above: that the former actually *use* the diagram to make the inference, and the latter do not. For example, someone who knew CN2 and who was presented with line 9 could make a sentential inference to line 10 without ever having seen the diagram. But the same is not true of the inference using visual addition, as the name implies. Not only must the reasoner have seen the diagram before; she must actually use it if she is to make this kind of inference to the desired conclusion.

To sum up: the focus of this discussion has been on conscious reasoning; that is, on inferential transitions between conscious occurrent thoughts. In each of the inferences so far surveyed, it seemed that there was a valid conscious inference to the relevant conclusion. But in several places more than one type of inference is available; there are visual alternatives to substitutional inferences. And considering the phenomenology of the reasoning here suggests that, at least for some reasoners, such transitions are not substitutional and not linguistic. In the later discussion, we will want to prefer candidate theories of the apparent epistemic value of this reasoning that can explain these phenomena.

2.8 Prop. I.32 Reconsidered

I want to close this chapter by noting an interesting curiosity in Euclid's argument in Prop. I.32 that has, as far as I can tell, gone unremarked to date by his commentators. This is that, short though the argument is, it contains a logically superfluous step.

We can see this from the reconstruction. As it stands, the inference to line 16 is as follows:

(10) $\angle ACD + \angle ACB = \angle ABC + \angle BAC + \angle ACB$

...

(15) $\angle ACD + \angle ACB =$ two right angles

---

(16) $\angle ABC + \angle BAC + \angle ACB =$ two right angles [10, 15: by substitution]

However, the inference to line 16 need not use line 10. Rather, it could take line 9 instead as the first premiss, operating again by substitution.

(9) $\angle ACD = \angle ABC + \angle BAC$.

In this case, the inference would not use the claim of the equality of the angles on a straight line and the interior angles of a triangle; rather, it would draw directly on the decomposition of $\angle ACD$.

In Euclid's argument the equivalent to line 10 is line (IV), and this too can be dropped. Thus the reasoning may go from:

(III) But the angle ACE was also proved equal to the angle BAC; therefore the whole angle ACD is equal to the two interior and opposite angles BAC, ABC.

directly to:

(V) But the angles ACD, ACB are equal to two right angles; therefore the angles ABC, ACB, BAC are also equal to two right angles.

Of course, the unnecessary inclusion of (IV) may be an oversight. But whether it is or not, we can ask whether this detour actually achieves anything.

The answer depends on what, if anything, new is established by line (IV) of the Euclid's argument. Line 10, its equivalent here, goes beyond line 9 in bringing out what proves, given lines 12-14, to be the crucial relationship between the internal angles of a triangle and the angles on a straight line. This is not either of the conclusions for which the Euclidean Presentation argues. But it is of obvious importance to a proper understanding of the underlying geometry. It would not be too much to claim that someone who did not grasp this relationship, but who claimed to have even a moderate such understanding, would be mistaken.

If this is correct, it suggests that Euclid's argument has the important effect of explicitly bringing to the reasoner's attention the relations between the internal angles of a triangle and the angles on a straight line. An argument that omitted line 10 and contained an inference directly from line 9, together with line 15, to line 16 would not at any stage explicitly represent that relationship. Someone could in principle follow such an argument and yet fail to notice a fact that is central to a genuine understanding of the underlying geometry. Including line (IV), which makes the general relationship explicit, closes this possible gap in the reasoner's understanding. Such a closure comes at a small but detectable cost, since line (IV) is, strictly speaking, logically superfluous to the overall conclusion to be established. But its inclusion is a source, I would suggest, of some of the clarity noted earlier as produced by following this argument.

2.9 Summary

Let me sum up. This chapter has introduced, situated and analyzed Euclid's argument in detail. In the course of this analysis, it identified several places in which two or more different types of valid inference are available to the reasoner; and it drew a general contrast between substitutional inferences using sentences, and visual inferences using the diagram. Among the latter, three different visual inference-types were noted: inference by 'seeing as', inference by visual translation, and inference by visual addition. These courses of reasoning—substitutional and visual—all count as ways to follow Euclid's argument. However, in at least one case, I suggested that the visual inference better described the phenomenology of a careful reasoner's actual experience in following Euclid's argument. Finally, I identified a logically superfluous step in the argument.

This discussion has deliberately been small-scale and detailed. In the next chapters, I turn to some of the more general underlying epistemological issues at stake. The strategy of the discussion is to take the findings of this chapter as data, and to ask how well each of four candidate theories can explain and situate the kinds of reasoning described here, and address—or answer—various questions that arise.

3

The Framework of Alternatives

3.1 Introduction

The last chapter analyzed different kinds of reasoning that a reasoner could use to follow Prop. I.32. It tried to do so in a philosophically neutral way: so as to raise, while leaving open, the various epistemological questions with which we will be concerned henceforth. This chapter now sets out a general framework within which we can categorize various candidate theories that seek to offer epistemological accounts of these kinds of reasoning.

Many people, philosophers and non-philosophers alike, have believed that Euclid's argument succeeds; that is, that a reasoner can rationally come to believe, and perhaps to know, the angle sum law by following the argument of Prop. I.32. Others have disagreed, for reasons some of which were briefly surveyed in Chapter 1. This raises the central epistemological question with which this book is concerned: whether Euclid's argument can confer justification, and if so, how we should understand that justification. In particular, is the purported justification empirical or *a priori*? And: does the diagram contribute to the purported justification?

3.2 Two Questions about Justification

In order to address the overall question of justification, it will be helpful to start with the various candidate theories that have been advanced. We can, I will argue, categorize such theories by their responses to the latter two questions above. However, a preliminary question concerns what we mean at all by a justification's being empirical or *a priori*, and such that the diagram does or does not contribute to it. I want to leave open the possibility that different theories may employ slightly different notions under these titles in each case. But a general characterization will be enough at this point to allow us to make the relevant classifications.

Contributing to Justification

Take an interpreted sentential presentation of a given argument. We can identify at least three different kinds of function that the various sentences here might in principle have, for a reasoner seeking to follow the argument. In the first place, they might serve to *fix the argument*; that is, to establish in the reasoner's mind just what the argument is, or is about. Secondly, they might serve to *express claims*, and in particular propositional claims. Thirdly, they might serve to *mediate inference*; that is, to present claims in such a form that a reasoner can use the sentences themselves to infer further claims. And within a given presentation, various sentences might fulfill more than one function.

It is plausible that each of these functions must be adequately discharged for the reasoner to acquire justified belief, or knowledge, from following the argument; they are, that is to say, *epistemically necessary* for her to follow the argument with justification. If it is sufficiently unclear just what the argument is, then it will not be possible for a reasoner to get started at all. If the requisite claims are not presented, or if they are not presented in a form sufficiently suitable for inference, then it will not be possible for a reasoner to acquire justified belief or knowledge from following the argument. She may lack epistemically relevant information, or be unable to make certain epistemically relevant inferences.

Of these three functions, the first does not seem to bear on justification as such; it is more in the nature of a prerequisite, and the question whether a given argument justifies can only really be asked once it has been sufficiently established *what* argument is in question. The other two functions, however, clearly do bear on justification.

Something similar can be said for diagrams, in the context of diagrammatic or hybrid sentential/diagrammatic presentations of arguments. We can think of a diagram's contribution to justification in terms of what it distinctively offers to the justification conferred by a given type of reasoning; and in particular, by reference to how, if at all, it discharges the functions of presenting information and mediating inference. As with sentences, erasing a diagram from a given argument will—if the diagram is not superfluous, and if the erasure is not made good in some other way—have a negative epistemic effect on any reasoning required to follow that argument: either it may no longer be clear to a reasoner from the remainder of the presentation alone just what the argument is, or certain information essential to the argument may be lost, or certain inferential gaps or defects may appear, or a combination of these three things. But of these three, only the latter two bear on justification as such.

We can now broadly characterize what it is for a diagram to *contribute to justification* as follows. Take a process of reasoning by which a thinker can come to believe a given claim with justification. Then we can say that if that reasoning uses a diagram, and would be impossible, or would fail to justify (unless otherwise supplemented), in the absence of the diagram (or something similar), then the diagram contributes to the justification conferred by the reasoning in question.

Though it is imprecise, this captures the two main ideas we need. First, the reasoning with the diagram is, by hypothesis, sufficient to justify a thinker's belief in the relevant claim. Secondly, the diagram has positive epistemic value: drop the diagram without compensating for the loss in some other way, and the reasoning in question ceases to justify a subject's belief, or is impossible. Note that it is consistent with this general characterization that a diagram might offer some degree of justification within a given type of reasoning, but not enough to warrant the relevant belief; it would not contribute to justification in the sense identified above. Also, note that there is no claim here that a given type of reasoning with the diagram may contain only reasoning with a diagram; there is nothing to rule out its also containing sentential reasoning, for example. And there is no claim that there is only one type of reasoning that could constitute following a given argument with justification.

Applying this general characterization to the present case, we can say this: it needs to be shown that a thinker can come to believe the angle sum claim with justification by following Euclid's argument in a given way. If so, we can say that if reasoning in that way uses a diagram, and would be impossible, or would fail to justify (unless otherwise supplemented), in the absence of the diagram (or something similar), then the diagram contributes to the justification conferred by that reasoning.

As this brings out, we will want to be very sensitive in the discussion that follows as to what constitutes a way of following Euclid's argument. I suggested in the previous chapter that consideration of the relevant phenomenology could help us to individuate different processes of reasoning, at least *prima facie*. But we should note here that if this approach is not to be question-begging in relation to the issue of whether or not a diagram contributes to justification, specification of the relevant phenomenology cannot itself mention the diagram. It will not be of interest if a diagram is found to contribute to the justification conferred by a process that has already been specified as diagrammatic.

The A Priori vs. the Empirical

I also need to say what is meant here by the thought that a justification can be empirical or *a priori*. It has sometimes been noted that there have historically been two aspects to the idea of *a priori* justification:[56] a negative aspect (that a belief is justified *a priori* if it is justified independently of experience), and a positive aspect (that a belief is justified *a priori* if it is justified through the exercise of reason alone).[57] A belief is then empirical (or *a posteriori*) if its justification either is not independent of experience, or involves more or other than the exercise of reason alone. It would then be a further claim to equate the two aspects: that is, to claim that a belief's being justified independently of experience amounted to or was the same as its being justified through the exercise of reason alone.

In fact it is plausible that, as they stand, the negative claim is the weaker; for the positive claim that a belief is *a priori* if it is justified through the exercise of reason alone requires not only the claim that the exercise of reason justifies the proposition, but that it does so *alone*, i.e. independently of experience. (If this were true, the negative claim would be a presupposition of the positive claim.) In any case, for the purpose of this discussion the claims of reason as such are not at issue, and so it is the negative claim that is of interest.

The question then arises as to what is meant by 'experience'. In the last chapter, I briefly described what I called a broad conception of experience, but I now want to situate that by contrasting it with two narrower conceptions.[58] On the narrowest reading, 'experience' relates to sense-perception: the occurrent deliverances of the sense organs as to objects external to the subject's body. On a broader reading, it also includes the proprioceptive or kinaesthetic perception of the subject's own bodily states and events. On the broadest reading—that adopted in Chapter 2—'experience' includes not only perception of the external world and bodily states, but also the subject's awareness of conscious events and states of thinking, imagining and desiring.

Why should it be appropriate for the present discussion to adopt the widest reading? I suggested above that such a reading allowed us to describe a useful notion of the 'phenomenology' of a type of reasoning. But I also think there are other and more general grounds for preferring the widest reading. Recall that on the negative claim identified above, a belief is

---

[56] E.g. by BonJour 1998, Ch. 1.

[57] I am restricting attention to beliefs at this point; however, parallel remarks can be made about claims or propositions.

[58] On this cf. Boghossian and Peacocke 2000, Introduction.

justified *a priori* if it is justified independently of experience. The phrase 'independently of experience' is normally taken to mean something like 'in a way that is not epistemically reliant on experience'; that is to say, in a way in which the experience is not used as evidence for the claim, or for any member of a set of premisses from which the claim is inferred. In general, it will be true of the above classification that the wider the reading of 'experience', the more categories of states and events are available in principle to act as evidence, and so the narrower the resultant reading of the *a priori*.

Now there has been much general skepticism in recent years as to the status and nature of the *a priori*. So it is appropriate to assess claims that the justification derived from following Euclid's argument is *a priori* against the broadest reading of 'experience'. It might or might not then turn out to be the case that the claim that this justification is *a priori* can be made good; but if it can, this will not be so because the relevant conception of experience has been too narrowly drawn.

Finally, we need to ask if the two questions posed above—as to whether the purported justification here is empirical or *a priori*, and whether the diagram contributes to the purported justification—are logically independent of each other. If they were not, then a given answer to one might entail or be entailed by a given answer to the other. I do not think the answer to this is quite clear at this stage, given these very general characterizations. But a *prima facie* answer should be No. We did not use either notion in the course of characterizing the other, and there is no other evident logical link between them.

3.3 The Framework of Alternatives

With these distinctions in mind, we can now return to the classification mentioned above. Let us call the property of a justification's being such that a diagram contributes to it, the property of being 'diagrammatic'. There are then four logically possible alternatives. The theory may hold that the justification conferred by following I.32 is (i) empirical and diagrammatic, (ii) empirical and non-diagrammatic, (iii) *a priori* and non-diagrammatic, or (iv) *a priori* and diagrammatic.

The various alternatives can be classified in the form of a matrix, as follows:

		Does the diagram contribute to justification?	
		Yes	No
Is the justification *a priori*?	Yes	*A Priori I* Justification is *a priori* and the diagram contributes to justification	*A Priori II* Justification is *a priori* and the diagram does not contribute to justification
	No	*A Posteriori I* Justification is *a posteriori* and the diagram contributes to justification	*A Posteriori II* Justification is *a posteriori* and the diagram does not contribute to justification

Provided it can answer the two questions above in a yes/no way, any candidate explanation of this reasoning may be located within this matrix, which I shall henceforth call the Framework of Alternatives. Given this proviso, the framework exhausts the available alternatives.

The next four chapters explore candidate explanations of this reasoning, which I attribute to, respectively, a particular interpretation of Plato by Sir David Ross; J.S. Mill; Leibniz; and Kant. I shall argue that each of these occupies one of the positions identified in the Framework of Alternatives, as below:

		Does the diagram contribute to justification?	
		Yes	No
Is the justification *a priori*?	Yes	*Kant*	*Leibniz*
	No	*Ross's Plato*	*Mill*

These are, of course, not the only theories (if we can call them that) that have been advanced. And indeed they may not be the four best-known such theories; for that one might need to include Berkeley's views, aspects of which I shall also discuss in Chapter 7 below. But I shall argue that the four chosen accounts have the merits of being both historically well-founded and of allowing the various issues to be presented in a relatively clear way. And it is with these issues, and not the historical theories as such, that I shall be primarily concerned.

A Dilemma of Justification?

Given the broad conception of experience outlined above, much of the debate that follows will concern what we might term *the uses of experience*; and in particular the question whether, and if so in what sense, the reasoner's experience in following Euclid's argument serves as evidence for her belief. Here it may seem at first as though there are just two alternatives: either experience of the diagram serves as evidence for a reasoner's belief, in which case it contributes to justification of that belief; or it does not. If it does, then the reasoning is empirical. If it does not, one may think, the diagram does not contribute to the justification of the reasoner's belief.

Someone who believes that these are the only alternatives is committed to the bottom-left or top-right hand box in the Framework of Alternatives. Such a person might claim that there was, in effect, a dilemma of justification here: that a commitment to the reasoning's being *a priori* carried with it a commitment to the diagram making no justificatory contribution, and a commitment to its making such a contribution carried with it a commitment to the reasoning's being empirical.

The two questions illustrated by the Framework of Alternatives are, at least *prima facie*, logically independent ones, as we have seen. The dilemma therefore is a false one, logically speaking: two other positions are logically available. But are these positions epistemically possible? Can the justification be empirical and yet receive no contribution from the diagram? Can the diagram contribute to justification, but in an *a priori* way? To these positions I shall turn, in Chapters 5 and 7. The next chapter, however, explores the first type of candidate account—that of Ross's Plato—in more detail.

4

Crude Empiricism: Ross's Plato

4.1 Introduction

The last chapter described a Framework of Alternatives in terms of the responses to be given to two questions, relating to the justification apparently conferred by the type of visual reasoning discussed in Chapter 2: Is the justification *a priori*? Does the diagram contribute to justification?

The lower left-hand box of the Framework includes theories on which, as indicated, the justification is *a posteriori* (i.e. empirical) and the diagram contributes to justification. One such theory has been advanced by Sir David Ross in an interpretation of a passage in Plato. According to this view, in following Euclid's argument, a reasoner gathers evidence by means of sensory experience of the diagram, and the conclusion is reached by an inference from this experience.

The present chapter explores this view, and the considerations for and against it, in more detail. I will consider two related types of account. On the first, a reasoner is supposed to be able to infer the general conclusion just by following Euclid's argument once in relation to a single diagram. On the second, a reasoner infers the conclusion by generalizing from her experience of several diagrams. If we think of 'inductive inference' very broadly, as referring to inference from the particular to the general, we might consider both of these accounts inductive, the former being from a sample of one. The topic of inductive inference is large and complex, and I cannot hope to explore it in detail here. The strategy of this chapter is rather this: to bring out, albeit in a fairly crude and introductory way, the main considerations that work for and against the different accounts. Further considerations for and against another (slightly different type of) inductive view will emerge in the next chapter.

4.2 Ross on Plato

There is a well-known passage in Plato's dialogue the *Meno*, in which Socrates leads a slave-boy through a piece of geometrical reasoning, apparently using a diagram. The reasoning addresses the problem of how to construct a square that has exactly twice the area of a given square. It proceeds by constructing a second square, of sides equal to the diagonal of the first square, and then showing of the second square that it is composed of four isosceles triangles of area equal to that of the triangle formed by two sides and the diagonal of the original square. Since the area of the original square was twice that of such a triangle, and the area of the constructed square is four times that of the same triangle, the area of the constructed square is exactly twice that of the original square. The apparent state of affairs has been represented diagrammatically as follows:[59]

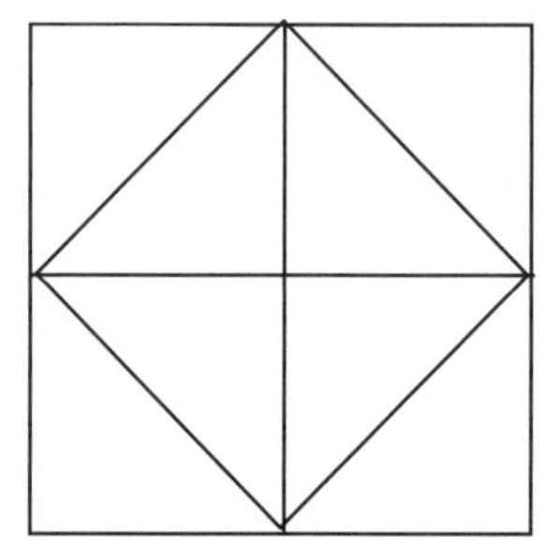

In a brief discussion of this passage, Sir David Ross remarks as follows:

> The method by which the slave-boy is got to discover what square has twice the area of a given square is a purely empirical one; it is on the evidence of his eyesight ... that he admits that the square on the diagonal of the given square is twice the size of the given square. He admits that certain triangles have areas equal, each of them, to half of the given square, and that the figure which they make up is itself a square, not because he sees that these things must be so, but because to the eye they look as if they were.[60]

---

[59] Sharples 1985, p. 69.

[60] Ross 1951, p. 18; my italics.

The problem of how to construct a square that has exactly twice the area of a given square does not, as far as I am aware, occur in Euclid. But it could readily be presented in Euclid's geometry using only the resources developed in Book I, and its conclusion is an instance of Pythagoras' theorem (Prop. I.47, where, however, the argument is quite different).[61]

4.3 Single-Diagram Induction

I shall not, however, discuss the reasoning in the *Meno* here as such.[62] It is fairly evident that Plato did not in fact hold the view that Ross describes above, and Ross himself may not have done so. But I will refer to this view as that of Ross's Plato for the sake of convenience. Ross's remarks do not make clear how the details of this view are to be filled out. But its general cast is clear enough from the italics above: the justification here is empirical, and the diagram provides the requisite evidence. So it is appropriate to classify it in the lower left hand box in the Framework above.

The question is whether the view of Ross's Plato or something similar could constitute an account of what it is to follow Euclid's argument with justification. Perhaps Ross is suggesting this: that following the argument in relation to a given diagram furnishes sufficient empirical evidence for a reasoner to be entitled to infer a general geometrical truth. This view can be motivated to some extent by noting that we can apparently follow arguments in circle geometry in which we only need to see a diagram once to know that the relevant conclusion holds generally. If the latter process is empirical, then maybe following Euclid's argument is like this.

I shall argue below that it is in fact not possible to follow Euclid's argument in the way required here. But even if we ignore this point, this first inductive view is, I suggest, unsustainable. Recall that Euclid's conclusion is not a claim about a particular diagram, but a claim about all triangles. So if the diagram is to justify belief in Euclid's conclusion, it must furnish empirical evidence that the angle sum claim holds for all triangles. But how can it do this? How can a single diagram furnish enough evidence for a reasoner rationally to reach this entirely general conclusion? What may secure some reasoning with a diagram in circle geometry to a conclusion about all circles is the known background fact that all (plane) circles have the same geometrical properties. But this is not true of triangles: different

---

[61] This is just because for the right-angled triangle formed by drawing the diagonal on a square of side unit length, the square on the hypotenuse will, according to Pythagoras' theorem, be 2 x (unit length^2).

[62] On the epistemology here, see especially Vlastos 1965 and Giaquinto 1993.

triangles can have different geometrical properties. Moreover, there are no circles to which a suitable diagram of a circle cannot be taken to have a visual resemblance. But a diagram can be used to represent triangles to which it has no visual resemblance, and these triangles can have different geometrical properties from those of the triangles that the diagram does visually resemble. So a reasoner cannot reliably infer the geometrical properties of triangles generally just from a visual experience of a single diagram. This creates a central puzzle for any candidate account of our target reasoning: how can the reasoner who follows Euclid's argument be justified in believing that the angle sum claim holds for *all* triangles?

4.4 Multiple Diagrams: The Inductive View

There is an obvious rejoinder to this line of thought. This is that the view of Ross's Plato ought to hold that the generalization here is by induction on *several* diagrams. Maybe it is like this: the reasoner uses the rubric to draw a particular diagram of a triangle. She follows the Demonstration and observes that, as far as she can tell, the diagram is triangular, and the internal angles of (the relevant portion of) the diagram sum to two right angles. She then constructs another such diagram, and observes that it too seems to have the same property. On this basis, she formulates a general hypothesis: that every triangle has the angle sum property. This hypothesis is then tested by her and others by a process of experimentation, in which she draws and examines further diagrams, or imagines certain similar figures. (Maybe she re-runs the argument for diagrams of equilateral, isosceles and scalene triangles.) The inference to the general conclusion is, then, via an inductive generalization from sense-experience.

This is more plausible: let's call it the Inductive View for convenience. The Inductive View seems to create a tight connection between the answers offered to the two questions above. Why should one think the diagram contributes to justification? On this view: because the diagram presents evidence for the reasoner's belief. Why think the justification is empirical? On this view: because it relates to the diagram, a physical object, and the visual information that the reasoner derives from the diagram. Indeed, as noted in the previous chapter, it might even seem as though the Inductive View is obligatory for someone who thinks that the diagram makes a justificatory contribution.

But here is a difficulty: the reasoning process just described goes well beyond the kinds of reasoning discussed in Chapter 2. There it was noted that a reasoner does not need to consider more than one diagram in order to derive apparent justification for her belief; where she does do so, moreover, it seems to be to confirm that she has implemented the construction process correctly, not to justify her belief in the conclusion of the argument. Moreover, in the kinds of reasoning described in Chapter 2 there is little that can be plausibly described as *evidence-gathering.* For example, it makes no difference whether the diagram has lines that seem or are perfectly straight or angles that seem or are perfectly sharp.[63] And the reasoner does not need to scrutinize the diagram carefully at the outset, or to monitor it during the course of the reasoning, to ensure that it or its properties have not changed. These considerations undermine the claim of the view of Ross's Plato to be an account of the type(s) of reasoning that we have taken as our target.

Such considerations are hardly conclusive: perhaps we were just mistaken about the phenomenology of the reasoning noted earlier, or the physical state of the diagram is so obvious as not to need checking. But they can be strengthened by considering some of the properties of the beliefs acquired as a result of following Euclid's argument. I noted in Chapter 2 that, among other things, the reasoning involved in following Euclid's argument appeared to give rise to feelings of accessibility and certainty in a reasoner. It seemed as though a reasoner who follows the argument can acquire a belief in the conclusion quickly; that is, that the transition from understanding the conclusion to believing it is a short one (possibly, for some inferences, even a phenomenologically immediate one). And it seemed as though that belief amounted to a strong conviction, a feeling that matters could not, or not easily, be otherwise.

Can the Inductive View explain these feelings of accessibility and certainty? Take the question of accessibility first. In many inductions to new belief, the reasoner does not make a rapid transition from understanding a claim to belief in it. Rather, after following an argument through for a few early cases, she entertains a general claim as a hypothesis; she then comes to form a (frequently tentative) belief in the hypothesis after further experimentation. There are at least two types of case, however, in which the transition to belief might be very rapid. In the first, a belief is formed via an inductive generalization and becomes entrenched and familiar. New data then emerge, where it is obvious that these fall under the existing hypothesis. The scope of the existing belief is slightly extended to cover the new data, and this process can be very rapid. A second type of case is one in which an entirely new belief is formed, but it is based on an overwhelming preponderance of evidence on one side, and little or no recalcitrant evidence on the other. Is our target reasoning one of either of these types of case?

---

[63] I use 'sharp' to describe the appearance of an angle where the component sides appear to overlap at exactly one point; an obtuse angle can be sharp in this sense.

Surely not. Belief in Euclid's conclusion is not a mere extension of existing belief, and there is no preponderance of evidence: quite the contrary.

Now take the question of certainty. Some inductive generalizations are such that, other things being equal, the strength of the belief produced is rationally positively related to the number of instances in the sample that constitutes the evidence. A belief that all sheep are woolly can thus rationally be strong for someone who has encountered a large number of woolly sheep. And some inductive generalizations are such that the degree of belief resulting from them grows incrementally with new (positive) evidence. The degree of belief may initially be low, especially in the early stages of experimentation, and so very sensitive to new evidence. The subsequent rate of growth in degree of belief may be large for further increments of (positive) evidence, and it may even start to decline, but it is not normally negative. But if there is an inductive generalization here, it is not like one of these. For the evidence in this case, the sample of beliefs reached by drawing different diagrams, is likely to be rather small. If it is not, it will not respect a fair description of what actually takes place; the reasoner does not, as we have seen, normally draw many different diagrams or visualize many different figures. If the belief here is generated by a generalization like those mentioned above, then given the paucity of evidence, the degree of belief generated should be relatively small. But in the case we are discussing it is not small; on the contrary, it typically amounts to a strong conviction, as noted. Moreover, it is not low at the outset. In this case, the degree of belief in Euclid's conclusion is equal or close to 1; this degree of belief is achieved very soon after the reasoner concludes the argument; and it shows little or no sensitivity to new 'evidence'.

Again, these considerations are not conclusive: maybe there are suitable models of induction that can explain these phenomena. But we can get a better understanding of what is at stake here by exploring the sharp contrast between the kinds of reasoning we saw in Chapter 2 and a process of mathematical reasoning that is clearly empirical. Imagine the reasoner who is unaware of the angle sum law and is asked to measure a diagram of a triangle with a protractor, as part of an effort to see if the internal angles of triangles always sum to 180°. Here we might well see all the factors mentioned above in play: very precise drawing of the diagram, careful attending to lines and angles, tentative formulation of the hypothesis, keen awareness of the possibility of error, the desire to draw many different diagrams to confirm the hypothesis, increasing confidence in the hypothesis, and perhaps final strong, but still conditional, belief. None of these needs to be in play in the kinds of reasoning surveyed in Chapter 2.

4.5 The Inductive View: Further Considerations

The Inductive View thus faces serious difficulties in providing a satisfactory explanation of our reasoning. And three further objections can be made, which I think make this view unsustainable. These derive from the requirement to take seriously the claim that, in this kind of reasoning, it is the reasoner's experience of the diagram that confers justification.

Explaining Euclid's Inferences

On the Inductive View, a reasoner can be empirically justified in reaching a certain belief by following Euclid's argument. The first question is whether it can explain how this occurs. On the Inductive View, the reasoner's visual experience of the diagram is supposed to provide evidence for her belief. Let us imagine then that she has very carefully drawn a diagram, including a triangle ABC, the extension of line BC to D and the auxiliary line CE, such that—so far as she can tell—the lines are perfectly straight and the angles perfectly sharp. She then proceeds to the first step of the Demonstration:

(I) Then, since AB is parallel to CE, and AC has fallen upon them, the alternate angles BAC, ACE are equal to one another. [I.29]

Say that she now acquires the belief that the alternate angles BAC, ACE are equal to one another. What justifies this belief? There seem to be three alternatives: either it is justified by the reasoner's application to the situation represented in the diagram of the rule given by Prop. I.29 (that alternate angles are equal), or it is justified by the reasoner's visual perception that, as far as she can tell, the two angles in the diagram she has drawn are in fact equal; or it is justified by both of the above.

None of these alternatives is acceptable, on the hypothesis that it is the experience of the diagram that provides justification. On the first, the experience of the diagram is not serving as the justifier; it is merely serving to represent an instance of a general rule that the reasoner is supposed antecedently to know. The justification is really being conferred by the reasoner's knowledge of the rule. But this is contrary to hypothesis. The second takes it that the experience of the diagram is the justifier; but then it is hard to see how the reasoner can be said to be following the argument at all. For the argument is explicit in describing an inference by application of a previously given rule, and the reconstruction given by the Inductive View makes no mention of this rule. So, thirdly: could both the rule and the experience of the diagram serve as concurrent justifications? Clearly not, for two reasons. First, if neither can serve as a justification, then both together cannot serve as concurrent justifications. Secondly, we can imagine circumstances in which the two could come apart from each other. Imagine that on re-inspecting the diagram, the reasoner sees that in fact, as she has drawn them, the two angles BAC, ACE are not quite exactly equal. If both the rule and the experience of the diagram are supposed to be able to justify, then this is a case in which they justify contradictory conclusions. In this case, the reasoner would have no justification at all.

It seems there is a quite general problem here for the Inductive View. For the same questions can be asked in relation to other inferences from the diagram in Euclid's argument, and in each case there is the same tension between the requirement for the experience of the diagram to be the conferrer of justification, and the constraint that the inference so construed be plausibly part of the kind(s) of reasoning involved in following Euclid's argument. The latter is a presupposition of this discussion; but maybe the former requirement could be relaxed? Maybe the view could be extended to allow that some inferences were permissible that were not on the basis of sense-experience? This would in itself be an important concession, but the question then is: which ones? We seem to have no basis for identifying some inferences as of one type rather than another. But say some such basis were found, such that in cases such as that above, the reasoner was required to adopt one type rather than another: would this solve the problem? I do not think so. For the reasoner would be disregarding an experience of the diagram that was in other circumstances sufficient to justify belief, or she would be disregarding the argument's instruction to apply a given rule in a given way. So the tension would remain.

Contrary Evidence?

Secondly, we need to consider more closely the apparent possibility and nature of contrary evidence provided by experience of the diagram. On the Inductive View, a reasoner would perhaps have contrary evidence if, after following the argument for some diagram, she had a sensory experience of the diagram in question on which the diagram did not exhibit the angle sum property, and she was satisfied that, so far as she could tell, the diagram had not been incorrectly drawn. If the internal angles added up to more than two right angles, for example, then this would be evidence against the conclusion; it would not be true that for any triangle, the internal angles of that triangle are equal to two right angles. This would be enough to disprove a universal generalization. If more of her evidence were like that, it might be enough to disprove a conditional, or perhaps probabilistic, generalization.

If all her evidence were like this, it would be enough on the Inductive View to support the contrary conclusion.

There are two points to be made here. The first is that, again, this does not describe how the reasoner normally reasons in following Euclid's argument. She does not take the existence of such a counter-example to undermine her belief in the general conclusion. Rather, she continues to hold that belief, and rejects the recalcitrant evidence. This is similar to the point about certainty made above, and suggests that she is making a deductive generalization. In generalizing from our experience of emus as flightless birds, we are normally prepared to admit that it remains, to greater or lesser extent, conjectural: there *might* turn out to be a flying emu, for example. In this case, however, the reasoner's belief in the conclusion of Euclid's argument seemed to be accompanied by the further conviction that there *could not* be a counterexample to it; that there just could not be a triangle of the relevant kind whose internal angles did not sum to two right angles.

Now, of course, the mere fact that we may be unwilling to give up a well-entrenched belief does not demonstrate apriority, for an empirical belief may be similarly well-entrenched. Nineteenth-century physicists did not, initially at least, give up their belief in Newtonian mechanics in the face of contrary experimental findings. Moreover, the fact that a counterexample may be inconceivable might simply be a reflection of our limited human cognitive capacities, and certainly does not rule out the possibility that we might learn to conceive one.[64]

Nevertheless—the second point—I would argue that the onus shifts here to the proponent of the Inductive View, to say what could constitute contrary evidence to the reasoner's belief. Could there be contrary evidence in this case at all? A reasoner could follow Euclid's argument and consistently believe that the diagram did not display the angle sum property, that no other diagrams that she drew displayed that property, and even that no diagram she could ever draw could display that property, and yet that Euclid's conclusion is true. But this implies that nothing could, even in principle, constitute evidence contrary to her belief. If so, the Inductive View is clearly mistaken. For if a reasoner is not even in principle able to reject Euclid's conclusion on the evidence of her experience of the diagram, then it cannot be claimed that her positive reason to accept it derives from such evidence.

---

[64] Mill 1843, Bk II, 5.6ff gives early form to this objection.

Reliability

Finally, we can revisit a difficulty raised in the previous section. The Inductive View holds that the reasoner is justified by empirical evidence in her belief in Euclid's conclusion. But again, Euclid's conclusion concerns triangles, not diagrams: it is the claim that all triangles have internal angles equal to two right angles. Could someone use inductive reasoning to acquire a justified belief in, or knowledge of, *this* claim, even in principle? It is hard to see how she could. It is plausible that it is a threshold requirement on something's being a justification that it should be reliable: that it (tend to) issue in true beliefs. It is unlikely, though not logically impossible, that there could be a geometrically perfect diagram. But say there were such a diagram or diagrams: it lies outside the limits of human perceptual capacities to detect the difference between such diagrams and some non-geometrically perfect diagrams that appear visually identical to them. So in generalizing inductively from a set of apparently perfect diagrams, a reasoner might—for all she knew—in fact be generalizing from non-perfect diagrams. But from *this* evidence the inference to the general conclusion would not be reliable. So, if reliability is a condition on justification, this evidence cannot constitute a justification for a belief in that conclusion. But if this is true, then someone cannot acquire a justified belief (or, knowledge, if knowledge is taken to be subject to the same requirement on reliability) in Euclid's conclusion by this means.

4.6 Summary

We can develop at least two views from consideration of some remarks of Ross on Plato. On the first view, the reasoner is supposed to infer the general conclusion from the empirical evidence of her sense-experience of a single diagram. On a second view, termed the Inductive View, she generalizes inductively from experience of several diagrams. I have argued that neither account offers a satisfactory explanation of the kinds of reasoning required to follow Euclid's argument, as described in Chapter 2; and that neither is sufficient to justify belief in Euclid's conclusion.

In the next chapter, I turn to a second possible type of explanation for our target reasoning.

5

Subtle Empiricism: J.S. Mill

5.1 Introduction

Let us return again to the Framework of Alternatives from Chapter 3. This was set out in terms of the possible responses to two questions, as on the next page.

The lower right-hand box of the grid includes theories on which, as indicated, the justification is *a posteriori* (i.e. empirical) and the diagram does not contribute to justification. Such a theory can be plausibly attributed to John Stuart Mill. According to this view, the reasoning involved in following Euclid's argument is deductive, but the basic axioms of Euclid's geometry are justified on inductive grounds. The justification that such reasoning confers is, thus, empirical. However, the experience of the diagram in Euclid's argument does not contribute to that justification.

We saw in the last chapter that it is highly implausible that the reasoning used to follow Euclid's argument involves an inductive generalization based on sensory experience of the diagram. Mill's view tries to avoid this weakness. In his view, the axioms are empirical and known by induction, but thereafter the reasoning is deductive. Thus the reasoning involved in following Prop. I.32 is deductive, serving to transmit justification from the premisses to the conclusion, without the diagram playing any role in justification.

The present chapter locates and explains Mill's view in more detail, and it assesses the plausibility of Mill's distinctive claim that the axioms and definitions of Euclid's geometry are empirical generalizations.

		Does the diagram contribute to justification? Yes	No
Is the justification *a priori*?	Yes	*A Priori I* Justification is *a priori* and the diagram contributes to justification	*A Priori II* Justification is *a priori* and the diagram does not contribute to justification
	No	*Ross's Plato*	*Mill*

5.2 Mill on Mathematical Justification

Mill's view of mathematical justification derives from a deep-seated hostility to the idea of intuition, as the following quotation from his *Autobiography* makes explicit:

> The notion that truths external to the mind may be known by intuition or consciousness, independently of observation and experience, is, I am persuaded, in these times, the great intellectual support of false doctrines and bad institutions. By the aid of this theory, every inveterate belief and every intense feeling, of which the origin is not remembered, is enabled to dispense with the obligation of justifying itself by reason, and is erected into its own all-sufficient voucher and justification. There never was such an instrument devised for consecrating all deep-seated prejudices. And the chief strength of this false philosophy in morals, politics, and religion, lies in the appeal which it is accustomed to make to the evidence of mathematics and of the cognate branches of physical science. To expel it from these, is to drive it from its stronghold.[65]

Mill states that a central goal of *A System of Logic* is to provide an empirical explanation of the apparent necessity of mathematics and logic. For our purposes, I will take it that Mill's position can be summarized in terms of six claims, set out below. I shall dwell a little here on exegesis, both to identify the claim in question and show that it can plausibly be attributed to

[65] Mill 1873, p. 145.

Mill, and because relatively little attention has been focused on Mill's interesting view of the role of diagrams in geometry.[66]

1. *Geometrical definitions and axioms are generalizations about physical objects.* 'Geometrical objects' as such do not exist according to Mill. He thinks geometrical claims are generalizations about physical objects: 'Since, then, neither in nature nor in the human mind, do there exist any objects exactly conforming to the definitions of geometry, while yet that science cannot be supposed to be conversant about non-entities; nothing remains but to consider geometry as conversant with such lines, angles etc. as really exist; and the definitions, as they are called, must be regarded as some of our first and most obvious generalizations concerning those natural objects.'[67]
2. *Geometrical axioms are inductively confirmed by observation of the physical world.* 'It remains to inquire, what is the ground of our belief in axioms—what is the evidence on which they rest? I answer, they are experimental truths; generalizations from observation. The proposition, Two straight lines cannot enclose a space … is an induction from the evidence of the senses.'[68] By 'axioms' Mill seems to mean, in relation to Euclid, the Common Notions, and perhaps the Postulates (e.g. he mentions Playfair's Axiom, an alternative to the Parallel Postulate).[69]
3. *Geometrical definitions are hypothetical.* Though both the axioms and the definitions of geometry are supposed to be empirical generalizations, Mill draws a contrast between them. While the axioms are supposed to be 'exactly true' or 'accurate', the definitions are 'hypothetical' or 'fictitious'.[70] By this Mill seems to mean that the definitions are nearly but not exactly true—'The hypothetical element is … the assumption that what is very nearly true is ex-

66 Indeed Mill's views on geometry have been much less explored than those on arithmetic, though Kim 1982 is an important contribution.

67 Mill 1843, II.V.1: 'There exist no points without magnitude; no lines without breadth, nor perfectly straight; no circles with their radii exactly equal.' References to Mill hereafter in this chapter are to Mill 1843.

68 II.V.4.

69 I give the Common Notions in the EP below. 'Playfair's Axiom' is, in the form quoted by Mill, the claim that two straight lines which intersect one another cannot both be parallel to one and the same straight line.

70 II.V.3; II.V.1, note.

actly so.'[71] The axioms, by contrast, are not hypothetical in this sense, according to Mill. What secures our reasoning with definitions, if they are inexact? Mill's answer is, in effect, that for most inferences the assumption of exactness is harmless; but that the reasoner must always be prepared to drop this assumption rather than draw an incorrect inference in reliance on it.[72]

4. *Inductive confirmation of axioms can be by perceiving diagrams.* Mill seems to have this mind when he says 'The truth of the axiom, Two straight lines cannot enclose a space, even if evident independently of experience, is also evident from experience.'[73] But he also believes that visualizing a figure can also count as an experiential route to knowledge of axioms: '[We can] make ... mental pictures of all possible combinations of lines and angles, which resemble the realities quite as well as any which we could make on paper.'[74]
5. *Inductive confirmation is sufficient for knowledge of axioms.* This is implied by the first quotation in (3) above; and by its continuation: 'Experimental proof crowds in upon us in such endless profusion ... that we should soon have stronger grounds for believing the axiom ... than we have for any of the general truths that we confessedly learn from experience.' From Mill's perspective, *a priori* accounts of mathematics are simply otiose. They are not required to account for mathematical knowledge, and should be dispensed with.[75]
6. *Geometrical theorems follow from axioms and definitions by deduction.* 'When, therefore, it is affirmed that the conclusions of geometry are necessary truths, the necessity consists in reality only in this, that they correctly follow from the suppositions from which they are deduced.'[76]

---

[71] II.V.1, note.

[72] II.V.2: 'Any hypothesis we make respecting an object, to facilitate our study of it, must not involve anything which is distinctly false, and repugnant to its real nature: we must not ascribe to the thing any property which it has not; our liberty extends only to exaggerating some of those which it has ... and suppressing others, under the indispensable obligation of restoring them whenever, and in as far as, their presence or absence would make any material difference in the truth of our conclusions.'

[73] And note also his remarks in II.V.6.

[74] II.V.5.

[75] II.V.4.

[76] II.V.1. The quotation continues: 'Those suppositions are so far from being necessary, that they are not even true; they purposely depart, more or less widely, from the truth.'

It might be thought that (6) is contradicted by Mill's famous assertion in II.V.1 to the effect that 'every step in the ratiocinations even of geometry is an act of induction.' But this conflict is only apparent. The axioms of geometry are known inductively, and this makes the conclusions ultimately inductive; but the inferences from a given set of premisses within a geometrical argument are deductive, proceeding by the application of (inductively) known rules or principles rather than by any induction on the diagrams within the arguments themselves.[77]

Mill's position, then, contains at least two moves designed to sidestep criticisms similar to those of Chapter 4. The first, noted earlier, relates to the challenge of how to explain the reasoner's apparent feeling of certainty. Mill acknowledges a general need for his explanation to be psychologically accurate, and criticizes competing accounts for failure on this account. But he insists, for reasons that are by now familiar, that the appearance of necessity is misleading; there is no necessity here, and so there is no objection for the inductive theorist to answer. The second comes in the Mill's acceptance, as described above, that the reasoning is deductive. In Chapter 4, I argued that given the hypothesis that it is the experience of the diagram that provides justification, the Inductive View was unable to explain how a reasoner can follow inferences in Euclid's argument with justification. It seemed as though either these inferences proceeded by induction, in which case they did not track the application of previously established rules within the argument; or they proceeded deductively by application of those rules, in which case there seemed to be no role for the diagram or figure, which was contrary to hypothesis. Mill's view tries to avoid this worry. According to this view, as it is understood here, these inferences are deductive, in the limited sense identified above. The diagram has no purported justificatory role within the argument, and so the possibility of overdetermination is avoided.

Now the fact that on Mill's view the diagram does not have an justificatory role here does not rule out the possibility in principle that it could constitute evidence; it may be simply be that any evidence it provides is too weak to justify. As noted, (4) above claims that axioms can be inductively confirmed by perceiving diagrams. But for the reasons given, I think we can take it that Mill's position is, in effect, that the diagram does not contribute to the justification derived from following Euclid's argument. It is,

[77] And Frege apparently also notices Mill's treatment of geometrical reasoning as deductive (Frege 1885, p. 24 fn. 1), with the characteristic comment: 'It is remarkable that Mill too (op. cit., Bk II, cap. VI, §4) seems to express this view. His sound sense, in fact, from time to time breaks through his prejudice in favor of the empirical.'

therefore, appropriate to locate Mill in the lower right hand box of the Framework of Alternatives, as above.

5.3 Mill's View: Initial Considerations

On at least two counts, then, Mill's view responds to and seeks to correct apparent weaknesses in the inductive view described in Chapter 4. What emerges is a position that is, if we take it literally, highly revisionary of traditional views about geometry. According to Mill, many (perhaps all) geometrical claims are never strictly speaking necessary.[78] Moreover, they never achieve absolute certainty, since inductively known claims are never absolutely certain. Finally, such geometrical claims are not even *true*: for they are not strictly true of physical objects, and there are no non-physical objects according to Mill. Indeed, if knowledge is factive (that is, only of truths), then Mill's view implies that we never have geometrical knowledge. So Mill's purported explanations of geometrical knowledge need to be understood with this general objection temporarily set to one side.

In relation to Prop. I.32, Mill's view is, then, that the angle sum claim is an empirical generalization, and so holds only approximately for all triangles. The reasoning in Euclid is deductive, and transfers empirical warrant from axioms and definitions to conclusions. But the axioms and definitions are empirical generalizations; they assert general matters of fact. The certainty of geometry derives from what Mill takes to be the overwhelming evidence we have for the truth of these generalizations. The exactness of geometry is illusory, and derives from the reasoner's taking the definitions to be exact when in fact they are not.

5.4 Euclid's Axioms as Empirical Generalizations

Are Euclid's axioms empirical generalizations? Many of the same general arguments against the Inductive View advanced in the last chapter will also apply here, so my discussion will be brief.

Let us focus on the specifics of Euclid's argument. I will take it that Mill includes Euclid's postulates under the heading of 'axioms'. One postulate and one axiom are used directly in Euclid's argument. The first is

---

[78] There are several subtleties in Mill's views here, which do not affect the present discussion. But it should be noted that some claims, e.g. conditionals of the form 'if P then P' etc., are avowedly both true and necessary even on Mill's view. How so? Mill draws a distinction between 'verbal' propositions and inferences and 'real' propositions and inferences, and denies apriority and necessity only to the real propositions and inferences; and Skorupski 1998 has plausibly argued that conditionals such as that above should be taken not as propositional but as formulating rules of inference, and as such analogous to verbal propositions for Mill.

Postulate 2, which is used for the first construction step, which in turn warrants line 12 of the reconstructed argument:

P2. Let it be postulated: to produce a finite straight line continuously in a straight line.

Say we read this as: 'For any given straight line segment s, it is possible to produce (i.e. extend) s continuously in a straight line.' Is this an empirical generalization? Surely not. First, taken as a summary of human experience, the claim is not true, even approximately. As noted, it is often not physically possible to extend a given line at all, let alone in a straight line or continuously. We can readily think of counterexamples involving very small, very large or highly inaccessible lines: think of a straight line located under ground. So we do not have the experiential base for an acceptable inductive generalization, as Mill's account requires.

Secondly, Mill insists that the empirical evidence for axioms is overwhelming, and that this explains their privileged epistemic status, and why many have taken them to be *a priori*.[79] But there is no such overwhelming evidence in play here. Indeed, it is interesting to note that on one influential tradition, due to Aristotle, what distinguishes postulates from hypotheses (and axioms) is precisely that in the case of postulates, the reasoner may have a prior contrary opinion.[80] Though it is not clear that Aristotle has Euclid in mind here, this interpretation 'seems to fit Euclid's Postulates fairly well', as Heath remarks.[81] And it hardly suggests the presence of overwhelming evidence; quite the contrary. This is not an *ad hominem* point: on an empirical account one should, with Mill, expect the evidence for axioms to be overwhelming.

The second axiom or postulate used in Euclid's argument is CN2, which—as I have reconstructed it—is used to reach line 10. The relevant inference moves from

(9) $\angle ACD = \angle ABC + \angle BAC$

to

(10) $\angle ACD + \angle ACB = \angle ABC + \angle BAC + \angle ACB$ [9: by CN2, adding $\angle ACB$ to both sides]

---

[79] II.V.4.

[80] Aristotle, *Posterior Analytics*, I. 6, 74 b 5 (quoted in Heath 1956, pp. 117-9).

[81] Heath 1956, p. 120.

Recall that CN2 states that 'If equals be added to equals, the wholes are equal.' Again, we can ask whether this axiom is an empirical generalization. Let us imagine an opponent who simply denies that the available evidence from sense-experience is sufficient for us to have knowledge of the general claim that constitutes CN2. We often add equals to equals without the results equaling each other, she might say: some examples are rabbits, clouds and drops of water. But this evidence is contrary to CN2. Given this contrary evidence, the totality of evidence is insufficient for knowledge of the axiom. So the axiom cannot be known by experience.

Now Mill can reply to this kind of worry. It is open to him to say, in effect, that like other empirical generalizations, CN2 holds conditionally, against a background of auxiliary assumptions. These assumptions have the effect of ruling out deviant cases such as those (rabbits, clouds, drops of water) listed above. Gillies makes this point clearly, in relation to Mill's view of arithmetic: 'The laws of arithmetic are empirical generalizations, but hold only under certain conditions, namely that the objects counted should not reproduce or coalesce etc. As with other empirical laws, these limiting conditions are tacitly assumed, without being explicitly stated.'[82] And, indeed, Mill can go on to say that if contrary evidence were to emerge to a given axiom, then the proper response would be to reformulate the generalization and/or the auxiliary assumptions, and this process would itself illustrate that the original generalization was an empirical one.

The effect of this move is to give up CN2, since it will need to be replaced with a complex conditional specifying the auxiliary assumptions under which it is true. But the deeper point is that this process of progressively more complex refinement is most implausible in relation to axioms. On Mill's account, the empirical evidence for axioms is supposed to be overwhelming. But this can hardly be so once the process of refinement has progressed even a short way.

5.5 Euclid's Definitions as Empirical Generalizations

Now let us turn to Euclid's definitions. Are these covertly hypothetical generalizations about empirical objects? Euclid's definitions, and the status of definitions in general in mathematics, have been controversial; but we need not engage with these controversies here. Given the similarities between Mill's view of axioms and his view of definitions, the overall line of argument here should be fairly evident; it will not be surprising that there are several evident sources of difficulty for Mill here too. I will focus here on just one.

---

[82] Gillies 1982, p. 26.

Recall that Mill's picture is one in which a definition is reached by generalizing from a set of empirical objects with a given property and then attending to that property to the exclusion of others. Now consider Euclid's definition of a triangle as, in effect, a three-sided rectilinear figure.[83] According to Mill, this is reached by generalizing about triangular physical objects and then attending solely to the property of triangularity that they apparently share. Since the definition thus reached will not be exactly true for any object that he takes to fall under it, Mill postulates a hypothesis or condition, whereby we 'feign' that the definition is exact.

The problem is that Euclid's definition of 'triangle' is most plausibly regarded as a stipulation: it is specifying in advance a rule as to how the term 'triangle' is to be used in the text that follows. As a stipulation, the definition cannot be confirmed or disconfirmed. Compare the definition: a 'tweezil' is a man with one arm born in New Guinea after 1954. What could confirm or disconfirm this stipulation? Since the definition cannot be confirmed or disconfirmed, no evidence could count for or against it. So it cannot be an empirical generalization, as Mill's view requires.

5.6 Summary

A second type of epistemological account of the kinds of reasoning involved in following Euclid's argument can be found in the writings of John Stuart Mill. According to Mill, the justification afforded by following Euclid's argument is empirical, and the diagram does not contribute to the justification. The axioms and definitions of geometry are inductive generalizations from human experience of physical objects. I have argued that Mill's view is not sufficient either to account for our target reasoning, or to justify belief in Euclid's conclusion.

---

[83] Strictly speaking, Definition 19 concerns 'trilaterals', not 'triangles'; but we can ignore this detail. Note that since a figure is defined by Euclid as 'that which is contained by any boundary or boundaries', we have no need to talk of a figure's being 'closed'.

6

Leibniz: The Denial of Epistemic Value

6.1 Introduction

So far, I have examined two types of account of the justification apparently conferred by the type(s) of reasoning involved in following the Euclidean Presentation. On both of these responses, the justification is deemed to be empirical; where they differ relates to whether or not the diagram contributes to the justification.

I now turn to a third type of possible response to the original question. This view differs from the two examined so far, in that it takes the justification in question to be *a priori*, not empirical. However, it shares with the view attributed in the last chapter to Mill, the claim that the diagram does not contribute to the justification.

This view can be located in the upper right-hand box of the Framework, and it can be plausibly attributed to Leibniz. I will not explore Leibniz's general views of mathematics or logic in any detail here. But it is possible to describe a view of Euclid's reasoning that is both recognizably Leibnizian and motivated by modern views of logic, and this will be the target of the present chapter.

We saw in the two previous chapters that it is implausible that the reasoning used to follow Euclid's argument is empirical, whether it is construed as evidentially based on sensory experience of the diagram, or in terms of deductive inference from inductively justified axioms and definitions. The Leibnizian view avoids the difficulties facing these empirical views, by taking the reasoning to be *a priori*.

The focus of this chapter, however, will be on the other Leibnizian claim: that the diagram does not contribute to justification. In fact, however, the Leibnizian view does not so much argue directly against the diagram's contributing to justification, as presuppose a somewhat different background picture of reasoning and justification in general. The effect of accepting this picture, which has been highly influential, is to make it seem compulsory that the diagram is irrelevant to the justification offered by this reasoning. Accordingly, I will sketch this general picture, discuss the claims and commitments of the Leibnizian view itself, and assess it specifically in relation to Euclid's argument. This will enable us to judge whether the Leibnizian is correct to reject any contribution to justification by the diagram.

6.2 Leibniz and the Leibnizian View

In the *New Essays*, Leibniz responds to Locke by offering his own view of Euclid's arguments, as follows:

> But I do not agree with what seems to be your view, that this kind of general certainty is provided in mathematics by 'particular demonstrations' concerning the diagram that has been drawn. You must understand that geometers do not derive their proofs from diagrams, though the expository approach makes it seem so. The cogency of the demonstration is independent of the diagram, whose only role is to make it easier to understand what is meant and to fix one's attention. It is universal propositions, i.e. definitions and axioms and theorems which have already been demonstrated, that make up the reasoning, and they would sustain it even if there were no diagram.[84]

Considered in relation to the argument in Prop. I.32, this passage contains the following claims:

1. The Euclidean Presentation expresses an argument consisting of universal propositions (definitions, axioms, theorems).
2. These universal propositions are what justify the conclusion of the argument.
3. The diagram has a psychological function, which is to help the reasoner to attend to and understand the argument better.
4. The diagram does not contribute to the justification offered by the argument.

[84] Leibniz 1765, p. 360.

5. The expository approach of the geometers makes it seem as though the diagram makes a justificatory contribution.

We might also note the suggestion in the first sentence that 'particular demonstrations' cannot justify, which hints at Leibniz's overall view that *a priori* truths are general and not justificationally reliant on particular instances.[85]

Note that Leibniz does not dispute that the Euclidean Presentation can justify; this is implied by the conjunction of (1) and (2). Moreover, we know from external evidence that Leibniz believed that all mathematical justification is 'independent of the testimony of the senses' and so *a priori*;[86] and (4) above makes clear his view that in Euclid's arguments the diagram is justificationally irrelevant to the proof. So Leibniz clearly belongs in the upper right-hand box in the Framework of Alternatives above. And it is also evident that he holds several of the views discussed in Chapter 1: that the role of the diagram is merely psychological (by (3) and (4)) and that the argument conveys the misleading impression that the diagram does contribute to the justification (by (4) and (5)).

We can now define a Leibnizian view of our target reasoning simply as one that holds, with Leibniz, the two claims identified above: that the justification is *a priori* and that the diagram does not contribute to justification. This view, and in particular the denial that diagrams can have epistemic value, has become very influential; indeed, I noted in Chapter 1 that it is now the orthodoxy among philosophers, including philosophers of logic and mathematics. Why should this be? There are three main reasons, I would suggest. First, Euclid's diagrams were and are regarded as intrinsically liable to mislead, and so as unreliable. Secondly, the 19th Century saw considerable technical successes in reconstructing (partially) diagrammatic presentations of arguments with purely sentential presentations. This seems to have raised the possibility in principle that all mathematical arguments whatever might be presentable using sentences (of arithmetic, or of a logical language); and the further thought that only thus could the rigor of mathematical reasoning be assured. As noted, Bertrand Russell seems to have held these (or very similar) views, at least in 1900-1902.

The third reason for the denial of epistemic value to diagrams lies in the 20th Century development of the study of formal systems, initially, after *Principia Mathematica*, and then in pursuit of Hilbert's Program. The study of strictly formal derivation systems as representations of systems of proof in various bodies of mathematics led directly to a widespread emphasis on

---

[85] Cf. e.g. Leibniz 1765, p. 50.

[86] Ibid.

such formal derivations as paradigms of proofs. From this formalistic perspective, the suggestion that the diagram has epistemic value is akin to a category mistake. Euclid's argument is one in which the purported justification exploits a reasoner's grasp of the representational content of the diagram, as conveying information about geometrical shapes. In a formal derivation, by contrast, every transition can be completely specified as a purely formal syntactic alteration. Transitions that depend on a reasoner's grasp of the representational content of a diagram are ruled out *ex ante*, since such transitions cannot be specified as purely formal syntactic alterations.

On such a view, then, there can be no epistemic value to the diagram, even in principle. This line of thought is, I suggest, tacitly presupposed in many modern views of Euclid's geometry. Here, for example, is a well-known statement about diagrams in geometry from Neil Tennant:

> It is now commonplace to observe that the diagram [sc. triangle ABC] ... is only an heuristic to prompt certain trains of inference; that it is dispensable as a proof-theoretic device; indeed, that it has no proper place in the proof as such. For the proof is a syntactic object consisting only of sentences arranged in a finite and inspectable array. One is cautioned, and corrected, about ... the mistake of assuming as given information that is true only of the triangle that one has happened to draw, but which could well be false of other triangles that one might equally well have drawn in its stead.[87]

In this case again the diagram is regarded as an heuristic aid that is irrelevant to justification, and hence 'merely psychological'; and the potential of the diagram to mislead is emphasized. However, the passage also gives a further motivation for the dismissal of the diagram, in claiming that diagrams are simply out of place in proofs. A proof is a 'syntactic object consisting only of sentences arranged in a finite and inspectable array'. There seem to be two thoughts here: first, that as *syntactic* objects, proofs cannot contain diagrams (an echo of the line of thought described above); and second, that as they are composed of *sentences*, proofs cannot contain diagrams.

6.3 Are Diagrams Out of Place in Proofs?

Let us take the former question first: do considerations of syntax by themselves give us sufficient reason to think that diagrams are out of place in proofs? Of course, simply replacing one or more steps in a given sentential

---

[87] Tennant 1986, p. 304.

proof with a diagram is likely to be a non-starter. But this is because, and only because, such a diagram falls outside the syntax and semantics of the relevant sentential system, and because suitable rules of inference in relation to the diagram have not been formulated. If these deficits can be supplied, it is not clear what reason there could be in principle to rule out diagrams *ex ante* from proofs on syntactic grounds.

I will return to this point below. But now take the second question: must proofs be composed only of sentences? If this were true, then there could not be such a thing as a diagrammatic proof, even in principle. But, as it stands, this latter claim is false. Indeed, there are logics in which proofs are largely composed of diagrams. For example, take the diagrammatic system known as the alpha Existential Graphs (EG), developed by C.S. Peirce. Here are diagrams in EG representing the standard truth functors:

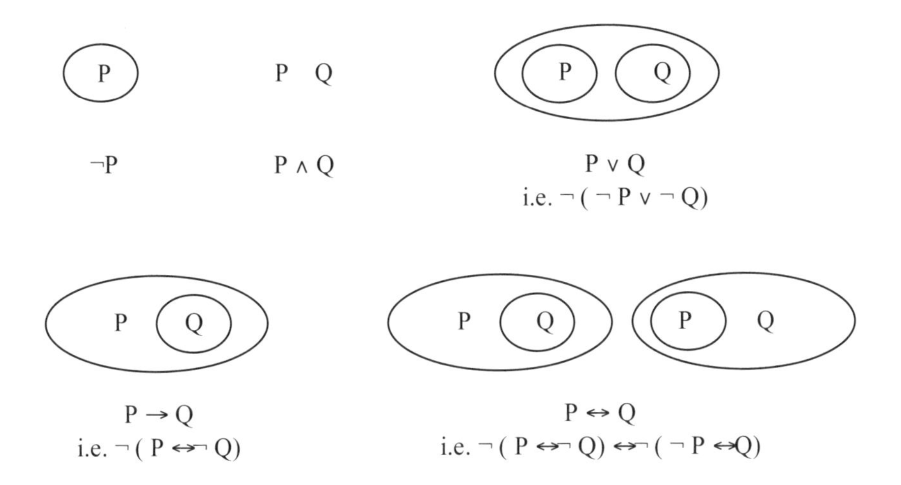

The vocabulary of the alpha graphs just consists of propositional letters and closed curves (known as 'cuts'). A letter by itself is taken assert a proposition; to enclose a letter with a cut is to negate it. There is no conjunction sign as such; two letters are conjoined by being written on a given region together. There is no need in EG for brackets, since the cut defines the scope of the relevant operator by what it encloses. According to the standard reading method given by Peirce, graphs can be translated into formulas of a sentential language by working inwards from the outside, reading cuts as negation and the open area between letters or subgraphs as conjunction. (The syntax, semantics and rules of inference for alpha EG are summarized in, for example, Roberts 1973 and Shin 2002).

Alpha EG is widely regarded as a diagrammatic system; and it meets certain other independent tests to be considered as diagrammatic.[88] Moreover, it has a specifiable syntax and rules of inference for the diagrams, and it is provably sound and complete.[89] In this system, however, every diagram is representationally homomorphic to a sentence in a standard two-functor ($\neg$, $\wedge$) sentence logic; that is, every mark in a diagram corresponds to a mark (or set of marks) in the relevant sentence in such a logic, a sentence that expresses the same proposition. So for any proof in alpha EG, an exactly parallel proof can be given in standard presentations of sentential logic by mechanically applying a rule translating the marks of each line of one into the marks of each line of the other. There is nothing to choose between alpha EG and standard propositional logic in point of rigor or logical expressiveness, though their representational forms are quite different.

So far, then, it is not true that diagrams are out of place in proofs. But an objector sympathetic to Tennant's view above might respond by drawing a distinction between what might be called intrinsically depictive and intrinsically non-depictive diagrams. Call a diagram *intrinsically depictive* if it can represent in virtue of a similarity of visual appearance with its object(s), *intrinsically non-depictive* if it cannot. Then Euclid's diagrams are intrinsically depictive, but the alpha graphs are intrinsically non-depictive: as noted, logical relations do not have a visual appearance, and so it is hard to see how they can be depicted by diagrams. So a revised objection might then be this: that intrinsically depictive diagrams are out of place in proofs. And the objector can motivate this latter claim by recalling Leibniz's own insistence that 'particular demonstrations cannot justify,' and claiming that it is the fact that they can depict that makes intrinsically depictive diagrams particular, and so problematic. This depictive quality is what connects such diagrams with a subject matter; it is what makes something a diagram *of a triangle*, for example. So again, it seems that what is suspect here is the fact that, in effect, presentations of arguments with intrinsically depictive diagrams improperly exploit their semantic value under a given interpretation.

This revised claim is more plausible. There clearly is the difference noted as regards depiction between the diagrams of alpha EG mentioned above and Euclid's diagrams. Moreover, we do have a precise syntax and semantics for the former, and not for the latter, as noted. And we can, and often do, reason with diagrams in alpha EG without regard to their semantics; but this is not possible for the diagrams in Euclid's arguments. For example, recall that in Prop. I.32 the rubric starts 'Let ABC be a triangle.' This is, in part, an instruction *to draw a triangle*; we cannot treat the word

---

88 On this see Norman 1999, Chs. 2 and 5.

89 As, e.g., shown by Roberts 1973.

'triangle' as a mark that could in principle be given different semantic values under various interpretations. So the revised claim seems to draw a genuine distinction between different types of diagram, and this distinction in turn seems to lend support to the underlying diagnosis that it is the appeal to the semantic value of diagrams, in Euclid and elsewhere, that is problematic.

Nevertheless, I suggest we should reject the revised claim. For there are various presentations of Euclidean geometry that use intrinsically depictive diagrams, and arguments using diagrams can be given in these systems that are clearly proofs. These systems work by defining what it is for a diagram to be well-formed, and then specifying suitable rules of inference for the system in question. This is the strategy adopted in, for example, Luengo 1996, which presents a diagrammatic subsystem that she claims (a) reconstructs a small fragment of Hilbert's (Euclidean) geometry, (b) has a specified syntax, semantics and rules of inference, (c) is provably sound, and yet in which (d) some (but not all) of the diagrams are intrinsically depictive. Miller 2001 presents a diagrammatic subsystem that he claims can be used to reconstruct most of the arguments of the early books of Euclid, and that also satisfies (b)-(d) above.

It is a mistake, then, to conclude that diagrams—whether intrinsically or non-intrinsically depictive—are out of place in proofs, or in mathematical or logical arguments more generally. As a result, Tennant's definition of proof as a 'syntactic object consisting only of sentences arranged in a finite and inspectable array' is needlessly restrictive.

6.4 The Commitment to Sentences

There is, then, no reason to hold a presumption in principle against diagrams in proofs, and I shall take it that the same holds *a fortiori* for arguments that are not proofs. Such a presumption would serve to rule out the possibility in advance that the diagram in Euclid's presentation contributes to justification. But to show that the presumption is mistaken is not to show that the Leibnizian view is wrong in this respect, and that the diagram does so contribute. To assess the latter question, we need to consider the specific role of the diagram here in discharging the two functions mentioned in Chapter 3 above: in presenting claims, and in mediating inference.

Before doing so, however, it may be helpful to make three preliminary comments. The first is just to remind ourselves that we can distinguish in general between belief states and propositions believed, and between the justification of belief states and that of propositions believed. A belief state is a type of mental state, which may take a proposition as content. The justification of a mathematical or logical proposition is in many cases a matter of its inferential relations to other propositions previously asserted or entertained in an argument. A belief state may be justified in many different ways: perhaps by following the relevant argument, but also, for example, by testimony. So the justification of a belief state may be in virtue of a reasoner's understanding of the justification offered by an argument of the claim believed; or justification may be achieved by some other means. Moreover, one can justifiably believe a proposition which is itself unjustified, as with someone who believes a mathematical falsehood on the basis of testimony from a previously reliable interlocutor. And one can unjustifiably believe a claim that it itself justified, as with someone who believes a mathematical theorem on the basis of the Tarot. What we are seeking here is a justification of a belief state, not of a proposition believed. We are not asking: 'What is the ultimate source of justification for the angle sum claim?' Rather, we are asking: 'What is the nature of the justification afforded to someone by following Euclid's argument in the way described earlier?'

Second, the effect of this is to clarify the nature of the explanatory burden on the Leibnizian View. There are, in effect, two tests to be met by the Leibnizian. First, the Leibnizian—who rejects any contribution by the diagram to justification—is committed to at least this opposing claim: that there is a way of following Euclid's argument with justification that does not rely on a diagram. He is committed, that is, to there being a purely sentential presentation of Euclid's argument. Call this the 'sentential presentation' test. But the mere existence of such a sentential presentation gives insufficient grounds for the Leibnizian to reject the diagram's putative justificatory contribution. So the Leibnizian must pass a further test: he must also show that following such a sentential presentation is a genuine way of following Euclid's argument as described in Chapter 2. Otherwise, he is vulnerable to the following objection: 'All you have given us is a sentential *counterpart* to Euclid's presentation; that is, an argument presented using sentences, which also argues to the angle sum claim. But you have given us no reason to think that the reasoning involved in following *this* presentation is the same as that we are seeking to assess.' In the absence of this further reason, the Leibnizian lacks the means to show the epistemic irrelevance of the diagram, as required. Call this the 'no counterpart' test.

Thirdly, we can make this more precise by recalling the notion of 'contributing to justification'. In Chapter 3 I suggested that if that part of a process of reasoning by which a reasoner arrives at a belief that P does not use a diagram (or similar) and is sufficient to justify the reasoner's belief state, then the diagram does not contribute to the justification of that belief state. This is necessary for the Leibnizian's claim to go through. But again, it is not sufficient; on pain of the 'counterpart' objection above; the Leibnizian must also show that the sentential presentation is a genuine way to follow Euclid's argument in the manner described. Without this, the Leibnizian is in fact in danger of showing not the irrelevance, but—contrary to his intention—the *indispensability* of the diagram to Euclid's argument; for the apparently diagrammatic thinking required to follow Euclid's argument with justification will still stand, and the Leibnizian's counterpart argument can simply be taken as capturing in an alternative sentential form the justificatory contribution made by the relevant thinking with the diagram.

6.5 Euclid's Argument Revisited: 1

With this in mind, let us examine Euclid's argument once again. I suggested in Chapter 3 that a given type of representation might contribute to justification in two different ways: first, by presenting a claim in an argument; and secondly, by mediating inference between one claim and another. We can see both these functions being discharged in the following inference:

(9) $\angle ACD = \angle ABC + \angle BAC$

---

(10) $\angle ACD + \angle ACB = \angle ABC + \angle BAC + \angle ACB$ [9: by CN2, adding $\angle ACB$ to both sides]

Here the sentences clearly have a presentational function. But the sentence in line 9 also mediates inference: that is, the reasoner can, just by manipulating it, reach the sentence in line 10. How so? The sentence in line 9 gives an equation of the form $X = Y$. Common Notion 2 (CN2) states 'Equals added to equals are equal.' So adding Z to either side of the equation yields an equation of the form: $X + Z = Y + Z$. Substituting back the relevant values for the variables yields line 10.

So far, then, the Leibnizian is on strong ground. This sentential inference does not require the diagram, either to present a claim or to mediate inference. So the Leibnizian has met the 'sentential presentation' test for this inference. But can he meet the 'no counterpart' test: i.e., can he show that this is a genuine way to follow Euclid's argument in the way described earlier? Fairly clearly, he can. Recall that the relevant part of Euclid's argument is as follows:

(IV) Let the angle ACB be added to each [sc. angles ACD; and angles BAC, ABC]; therefore the angles ACD, ACB are equal to the three angles ABC, ACB, BAC.

I presented a possible alternative visual route to this conclusion in Section 2.7. But it should be evident, given (IV) above, that the sentential inference is available—and perhaps even preferable—as a reconstruction of the thinking here.

So, as regards *this* inference, the Leibnizian has met both tests; it is not plausible that the diagram contributes to the justification afforded by the thinking involved in at least one possible way of following Euclid's argument at this point. Moreover, though every inference must be examined on its own merits, this overall line of attack is looking highly promising for the Leibnizian: there are several apparently similar substitutional inferences in Euclid's argument, as I reconstructed it in Chapter 2, such as those to lines 3, 6, 8, 9, 13, 15 and 16. If he could show that these inferences also met both the tests set out above, this would go a long way to establish his overall claim that the diagram makes no justificatory contribution to Euclid's argument.

6.6 Euclid's Argument Revisited: 2

But now take line 7, which is reached from the diagram:

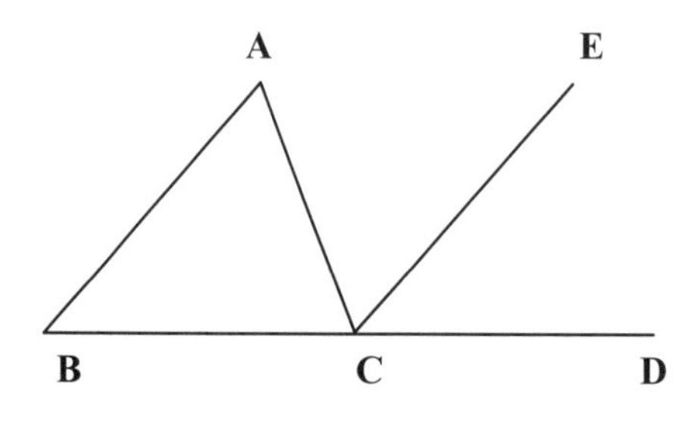

---

(7) $\angle ACD = \angle ECD + \angle ACE$

I reconstructed this inference in Chapter 2 as follows:

(7a) CE divides $\angle ACD$ into two parts, $\angle ECD$ and $\angle ACE$, without remainder [from the diagram]

(7b) The whole of an angle is equal in size to the sum of the sizes of any parts into which it is divided without remainder [background assumption]

(7c) $\angle ACD = \angle ECD + \angle ACE$ [7a, 7b: by substitution]

Thus reconstructed, this is a logically valid inference: lines 7a and 7b together entail line 7c (= line 7).

Again, this is a sentential presentation of the kind required if the Leibnizian is to show that the diagram is epistemically irrelevant to the reasoning required to follow Euclid's argument. But again, we need to recall the second 'no counterpart' test: is this reconstruction faithful to the thinking required to follow Euclid's argument? Fairly clearly, it is not. Note first that the reasoner does not seem to entertain the general thought in line 7b above, to the effect that the whole of an angle is equal in size to the sum of the sizes of any parts into which it is divided without remainder; and secondly, that the reasoner does not seem to do any substitutional reasoning to reach line 7—reasoning that would be required in using the general claim in 7b.

We can briefly sum up the position as follows. The Leibnizian cannot meet the 'no counterpart' test in relation to line 7 by this means; a very good candidate for the required sentential inference fails the second test. Moreover, it is very difficult to see how any purely sentential inference could fare better, since doubtless the diagram would still be required to present—and warrant—line 7a, or a similar premiss. It is, then, highly implausible that the route to justified belief here is via a sentential inference; rather, it seems to be via a piece of specifically geometrical thinking that uses the diagram. This part of the thinking involved in following Euclid's argument both uses a diagram and is sufficient to justify the reasoner's belief state. So, if we recall the discussion of 'contributing to justification' in Chapter 3, we can say that the diagram does contribute to the justification of that belief state here.

6.7 Euclid's Argument Revisited: 3

Is this an isolated result? No, for two reasons. First, there are other inferences in Euclid's argument that are similar to the inference to line 7 described above. The inference to line 11 is markedly similar, for example. This was reconstructed in Chapter 2 as follows:

(11a) AC divides the angle on line BCD into two parts, ∠ACD and ∠ACB, without remainder [from the diagram]

(11b) The whole of an angle is equal in size to the sum of the sizes of any parts into which it is divided without remainder [= 7b; background assumption]

(11c) ∠ACD + ∠ACB is the sum of the sizes of all the angles on BCD [11a, 11b: by substitution]

Again, the Leibnizian can point to a valid sentential presentation. But again, it is not plausible that this meets the 'no counterpart' test. And again, there is a geometrical, diagrammatic reconstruction that does so.

Secondly, I noted above that the logical inference to line 10 met both tests; it was both sentential and, plausibly, a genuine way to follow Euclid's argument. But we should not assume that other apparently similar sentential inferences are in fact genuine ways to follow the argument. Thus, take the inference to line 9:

(6) ∠ABC = ∠ECD [4, 5: by substitution]

...

(8) ∠ACD = ∠ECD + ∠BAC [3, 7: by substitution]

---

(9) ∠ACD = ∠ABC + ∠BAC [6, 8: by substitution]

Again, this reconstruction takes the form of an inference by substitution on sentences. (In Chapter 2 I described an alternative visual reconstruction of this thinking, in which the reasoner visually translates a copy of ∠ECD along line BCD until it is mapped on to ∠ABC.) Again, for familiar reasons, understanding this as a sentential inference does not seem to fit the phenomenology. But then it is again plausible that this is just a counterpart logical reconstruction of the relevant thinking here, and the true explanation is one that understands it as reasoning with the diagram, using a geometrical principle of translation of angles already known to be equal.

6.8 Epistemic Indispensability

What conclusions should we draw from the survey of inferences? Before answering this question, it may be helpful to draw a further distinction between two different ways in which a given type of representation—here, a diagram or sentence—might be said to contribute to justification. A representation *contributes* to the justification conferred by a given type of reasoning involved in following an argument, if that reasoning uses the representation in question, and would be impossible, or would fail to justify (unless otherwise supplemented) in the absence of the representation (or something similar)—this is just the sense of 'contribution to justification' already identified in Chapter 3. On the other hand, a representation *indispensably contributes* to the justification conferred by a given type of reasoning, if (a) it contributes to that justification and (b) there is no reasoning that does not employ such representations and that could genuinely be used to follow the argument.

Recall the inference to line 10 discussed earlier:

(9) $\angle ACD = \angle ABC + \angle BAC$

---

(10) $\angle ACD + \angle ACB = \angle ABC + \angle BAC + \angle ACB$ [9: by CN2, adding $\angle ACB$ to both sides]

Applying the distinction above, it is plausible to say here that the sentences indispensably contribute to the justification conferred by following this inference: the reasoning both uses sentences and is sufficient to justify the reasoner's belief state (contribution to justification); and it is hard to see how non-sentential—and specifically diagrammatic—reasoning could be used here to follow the argument as described in Chapter 2. (It may be helpful to recall again that the relevant line (IV) of Euclid's argument runs: 'Let the angle ACB be added to each [sc. angles ACD; and angles BAC, ABC]; therefore the angles ACD, ACB are equal to the three angles ABC, ACB, BAC.')

I will take it, then, that the distinction drawn here is fairly clear. With this in mind, we can ask whether the diagram indispensably contributes to justification in inferences such as that to line 7 described above. If the conclusion of Section 6.6 is correct, then the diagram certainly contributes to justification. But does it do so indispensably? Could it be true that there is no non-diagrammatic reasoning that could be used to follow Euclid's argument here as described in Chapter 2?

I do not think we can answer this question with absolute confidence either way: for all we know at present, there *might* perhaps prove to be a non-diagrammatic way to follow this argument in the way described. But, recalling the discussion at the end of Section 6.6, I suggest that the claim is highly plausible.

6.9 Summary

This chapter has argued that we can identify a Leibnizian View of the reasoning involved in following Euclid's argument, and that this view has been and remains highly influential. This view distinctively holds that the justification conferred by following Euclid's argument is *a priori*, and receives no justificatory contribution from the diagram; rather, the reasoning is sentential. In response, I have argued that if the Leibnizian View is to succeed, then it must be able to show: first, that there is a sentential presentation of each of the inferences in Euclid's argument; and secondly, that following the argument in this way is faithful to the reasoning described in Chapter 2. Applying these tests, however, implies that though it is correct to understand some parts of the thinking required to follow Euclid's argument as using sentential inferences, other parts use diagrammatic inferences. Someone who follows Euclid's proof in the way described in Chapter 2 uses both sentential and diagrammatic inferences. And in the latter, the diagram contributes to justification.

7

Kant: A Proto-Theory of Geometrical Reasoning

7.1 Introduction

I turn now to the fourth and—the reader may be relieved to learn—last category of possible response to our original question. This view differs from the three examined so far, in that it takes the justification in question to be *a priori*, not empirical, and yet such that the diagram contributes to justification.

This view can be located in the upper left-hand box of the Framework below, and it can be plausibly attributed to Kant; I will call it the Kantian View. Again, it is possible to differentiate between the Kantian View in general, and the specific cluster of views about geometry that Kant himself actually held. As with the discussion of Mill in Chapter 5, however, I will spend some time here on exegesis, since I think that—at least in relation to our target reasoning—Kant's views have been widely misunderstood. Since the epistemological account that I shall argue positively for in later chapters is recognizably Kantian, this emphasis is an appropriate one.

The Kantian View avoids the difficulties facing competing views according to the Framework of Alternatives, as discussed in Chapters 3-6. On the basis of the discussion so far, I will take it that this view correctly accepts that the diagram makes a justificatory contribution to the reasoning involved in following Euclid's argument (contra Mill's View and the Leibnizian View), while also correctly claiming that this contribution is not empirical and evidential (contra Mill's View and the view of Ross's Plato).

		Does the diagram contribute to justification?	
		Yes	No
Is the justification *a priori*?	Yes	*Kant*	*Leibniz*
	No	*Ross's Plato*	*Mill*

However, the combination of claims that constitutes the Kantian View has seemed to be a very unhappy one to many philosophers, and many objections have been raised against it. Behind these specific worries, there have been more general concerns. In what sense, if any, can a visual experience make a justificatory contribution without being evidential, thereby making the justification *a posteriori*? In what sense, if any, can an appeal to 'intuition' be part of a satisfying explanation of this type of reasoning?

A plausible positive account should, I think, not only address the specific worries; it should also endeavor to answer or defuse the general concerns that motivate such worries. What is required—and what I will seek to provide in the following three chapters—is a positive explanation and defense of (one version of) the Kantian View. First, however, I want to introduce and explore Kant's own account. Accordingly, the purpose of this chapter is: first, to show that Kant held what I have described as the Kantian View, and to outline Kant's response to the well-known Generality Objection; second, to argue that we can find in the *Critique of Pure Reason* a plausible though embryonic account of the kind of visual reasoning under consideration; and thirdly, to suggest that this account is at odds with an influential strand of interpretation of Kant, which may therefore require further review.

7.2 Berkeley and the 'Generality Objection'

We can introduce Kant's view by considering a central objection to any account on which reasoning with a diagram is taken to justify. The Generality Objection is simply this: How can the reasoner who follows Euclid's argument be justified in believing that the angle sum claim holds for *all* triangles?

Berkeley advances this point in the form of an imagined counter to his claim that we cannot have abstract general ideas, in the Introduction to the *Principles of Human Knowledge*:

> But here it will be demanded, how we can know any proposition to be true of all particular triangles, except we have first seen it demonstrated of the abstract idea of a triangle which equally agrees to all? For, because a property may be demonstrated to agree to some one particular triangle, it will not thence follow that it equally belongs to any other triangle, which in all respects is not the same with it. For example, having demonstrated that the three angles of an isosceles rectangular triangle are equal to two right ones, I cannot therefore conclude this affection agrees to all other triangles which have neither a right angle nor two equal sides. It seems therefore that, to be certain this proposition is universally true, we must either make a particular demonstration for every particular triangle, which is impossible, or once for all demonstrate it of the abstract idea of a triangle, in which all the particulars do indifferently partake and by which they are all equally represented.[90]

It will be recalled that in Section 13 Berkeley famously (and perhaps unfairly)[91] attacked Locke's claim that we can form general ideas in geometry, such as a general idea of a triangle that is (in Locke's words) 'neither oblique nor rectangle, equilateral, equicrural nor scalenon, but all and none of these at once'. As Berkeley notes, this rejection creates an apparent difficulty for his own view of geometry, for he concurs in the view that geometrical theorems are quite general ('universally true'), and if geometrical arguments do not employ abstract ideas, then it is unclear how they may be general at all. In particular, there seems to be no means in principle to prevent the reasoner from erroneously over-generalizing properties from the diagram that relevant triangles may not possess.[92]

Berkeley's response is as follows:

> To which I answer, that, though the idea I have in view whilst I make the demonstration be, for instance, that of an isosceles rectangular triangle whose sides are of a determinate length, I may nevertheless be certain it extends to all other rectilinear triangles, of what sort or bigness soever. And that because neither the right angle, nor the equality, nor determinate length of the sides are at all concerned in the demonstration. It is true the diagram I have in view includes all these particulars, but then there is not the least mention made of them in the proof of the proposition. It is not

---

[90] Berkeley 1988, Introduction, Section 16.

[91] On Locke here, see Ayers 1991, Chs. 5 and 27.

[92] Note that a reasoner might also *over-restrict* from the diagram: that is, from following the argument in relation to a diagram of a rectangular isosceles triangle, she might conclude that all rectangular isosceles triangles had the angle sum property. This would, strictly speaking, be an inaccuracy, but not a dangerous one; she would not have gone wrong by so inferring.

> said the three angles are equal to two right ones, because one of them is a right angle, or because the sides comprehending it are of the same length. Which sufficiently shows that the right angle might have been oblique, and the sides unequal, and for all that the demonstration have held good. And for this reason it is that I conclude that to be true of any obliquangular or scalenon which I had demonstrated of a particular right-angled equicrural triangle, and not because I demonstrated the proposition of the abstract idea of a triangle.[93]

Here Berkeley's point seems to be this. Imagine someone who follows Euclid's argument in relation to a diagram of a right-angled isosceles triangle, perhaps as below:

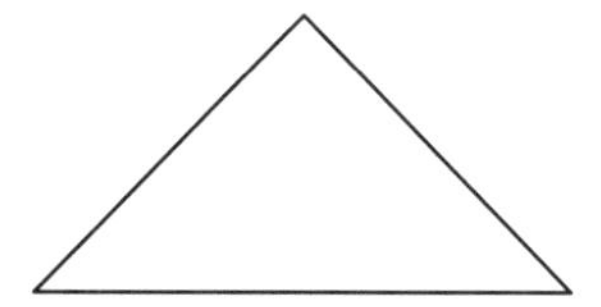

For such a reasoner, there is no valid generalization available merely from a visual experience of this diagram to a claim about all triangles, i.e. including all those that are neither right-angled nor isosceles, for reasons noted in Chapter 4. However, though the diagram above is of a right-angled isosceles triangle, these particular properties 'are not mentioned in the proof of the proposition'. That is, it is only *qua triangle* that it plays any role in Euclid's argument.

Berkeley says that the argument 'does not mention' the specific properties of being right-angled and isosceles, but this is insufficient, since an argument may fail to mention a property or claim and yet still implicitly rely on it. But bearing this in mind, his point is surely that Euclid's argument may be understood as justifying the angle sum property generally, because it does not rely on any claim about the triangle represented that is not a general property of all triangles. By contrast, the reasoner who, at the conclusion of the reasoning, takes the concluding claim only to apply to those triangles that the diagram visually resembles, has failed to follow Euclid's argument correctly.

Now this line of thought is hardly original to Berkeley: it can be found, for example, in Proclus.[94] It offers an intriguing and potentially plausible

---

[93] Berkeley 1988, Introduction, Section 16.

[94] Proclus, p. 207. On Berkeley, see Jesseph 1993, Ch. 1.

account of how the Generality Objection can be met in relation to Euclid's argument.[95] However, Berkeley's account itself might be thought to face a serious difficulty here.[96] For his account is vague at the crucial point, as to what if anything determines the representational scope of the diagram. There seem to be three candidate answers in principle here. The first is that it is the visual features of the diagram alone that determine its scope; but Berkeley rejects this fallacious appeal to the diagram, as we have noted. The second is that the diagram's scope is determined by the intentions of the reasoner, and these might in principle extend to triangles that the diagram did not visually resemble. But this is patently insufficient, since such intentions may vary from reasoner to reasoner; what is required is some independent yardstick that can be reliably used to determine scope by any suitably informed reasoner. The third is that the scope of the diagram is determined by the construction procedure specified in the text of the argument. But then it starts to seem as though Berkeley is tacitly appealing to an abstract general idea of a triangle. For it is quite unclear in virtue of what, if not abstract ideas, a reasoner may grasp *ex ante* the generality of a given construction procedure.

7.3 Kant's View

We can read Kant as developing a response of the third kind to the Generality Objection. Kant specifically discusses Euclid's argument in Prop. I.32 at the end of the first Critique, in the Transcendental Doctrine of Method. There he draws a general contrast between philosophical and mathematical cognition:

> **Philosophical cognition** is **rational cognition** from concepts, mathematical cognition that from the **construction** of concepts. But to **construct** a concept means to exhibit *a priori* the intuition corresponding to it.[97]

[95] But Berkeley's view is much less persuasive as an explanation of other general ideas. See Prinz 2002, p. 29.

[96] See Jesseph 1993, Chapter 3, and Coffa 1991, p. 46. I am also grateful to Marcus Giaquinto for helping me to understand the force of this objection.

[97] Kant 1998, A713/B741. Note that the recent Cambridge Edition translation (Kant 1998) denotes Kant's emphasis in boldface, reserving italics for foreign words as indicated in the originals.

A geometrical illustration immediately follows:

> Thus I construct a triangle by exhibiting an object corresponding to this concept, either through mere imagination, in pure intuition, or on paper, in empirical intuition, but in both cases completely *a priori*, without having had to borrow the pattern from it from any experience. The individual drawn figure is empirical, and nevertheless serves to express the concept without damage to its universality, for in the case of this empirical intuition we have taken account only of the action of constructing the concept, to which many determinations, e.g., those of the magnitude of the sides and the angles, are entirely indifferent, and thus we have abstracted from these differences, which do not alter the concept of the triangle.[98]

I interpret Kant's view here to be this: that the reasoner 'constructs' a concept of a triangle by drawing or visualizing a triangle. However, in doing so she is merely implementing the relevant construction procedure for triangles (i.e., has 'taken account only of the action of constructing the concept'). Although the drawn diagram is determinate in respect of the size of its sides and angles, the reasoner does not take these properties of the diagram to restrict the class of triangles that she takes the diagram to represent.

This interpretation is supported by some of Kant's later remarks, in elaborating his contrast between philosophical and mathematical reasoning:

> The former confines itself solely to general concepts, the latter cannot do anything with the mere concepts, but hurries immediately to intuition, in which it considers the concept *in concreto*, although not empirically, but rather solely as one which it has exhibited *a priori*, i.e., constructed, and in which that which follows from the general conditions of the construction must also hold generally of the object of the constructed concept.[99]

Again, here—to consider this quotation in relation to geometry—we find Kant emphasizing, first, the reasoner's ability to take an individual diagram to represent a general class of (geometrical) objects; and second and more explicitly than above, the role of construction procedures in determining what the diagram is to be taken to represent.[100]

---

[98] A713-4/B741-2.

[99] A715-6/B743-4.

[100] In what follows I will use the now-familiar term 'geometrical object' for explanatory purposes. But this is not to suggest that Kant believed that geometrical objects as such exist. Space does not permit discussion of this question, or of Kant's use of the key terms 'Gegenstand' and 'Objekt'; and Kant certainly talks of mathematical objects in places (e.g. B147). On this question, cf. the discussions in Parsons 1983, p. 147, and Friedman 1992, p. 94. Smit 2000 has an interesting and useful discussion of the terms above.

That it is Euclid's geometry that Kant has specifically in mind here is clear:

> Give a philosopher the concept of a triangle, and let him try to find out in his way how the sum of its angles might be related to a right angle. He has nothing but the concept of a figure enclosed by three straight lines, and in it the concept of equally many angles. Now he may reflect on this concept as long as he wants, yet he will never produce anything new. He can analyze and make distinct the concept of a straight line, or of an angle, or of the number three, but he will not come upon any other properties that do not already lie in these concepts. But now let the geometer take up this question. He at once begins to construct a triangle. Since he knows that two right angles together are exactly equal to all of the adjacent angles that can be drawn at one point on a straight line, he extends one side of his triangle, and obtains two adjacent angles that together are equal to two right ones. Now he divides the external one of these angles by drawing a line parallel to the opposite side of the triangle, and sees that here there arises an external adjacent angle which is equal to an internal angle, etc. In such a way, through a chain of inferences that is always guided by intuition, he arrives at a fully illuminating and at the same time general solution of the question.[101]

It is standardly believed that Kant is referring to Prop. I.32 in this passage, so that here we have an actual worked example which Kant uses to illustrate his general claims about mathematical cognition.[102] More specifically, however, this example is clearly intended to illustrate the claims discussed above: that an individual diagram may be used to represent a general class of objects; and that construction procedures in Euclid determine the scope of the diagram. However, the example also introduces two new ideas: first, by focusing on a case of *reasoning*, Kant implies that the inferences in Euclid's argument are valid ones; and second, he claims that intuition somehow has a guiding role.

Now Kant's focus in this last passage is on discovery of a geometrical truth, not on the process of following an existing argument as such. But his overall point surely applies to the latter: he is using Euclid's well-known argument to illustrate the epistemic necessity of intuition to a type of mathematical reasoning. On Kant's view, following Euclid's arguments is a paradigm route to geometrical knowledge; and geometrical knowledge

---

[101] Kant 1998, A715-7/B743-5; emphasis added.

[102] E.g. by the Editors of the Cambridge Edition (Kant 1998), p. 751.

would not be possible in the absence of a faculty of intuition. In particular, intuition of a diagram does not merely serve to fix the argument for the reasoner. Rather, it is taken to be an epistemic prerequisite: a condition on the possibility of the reasoner's having geometrical knowledge at all. So I think we can take it that for Kant the diagram contributes to justification in the sense described in Section 3.2. Mathematical knowledge for Kant is *a priori*. So, overall, Kant holds what I have termed the Kantian view, and it is appropriate to place Kant in the upper left hand box of the Framework of Alternatives, as above.

7.4 Kant on Construction in Pure Intuition

We can now return to the question posed in the previous section: What determines the representational scope of the diagram in Euclid's argument? It should be clear that Kant takes the third alternative, i.e. he claims that the scope of the diagram is determined by the construction procedure specified in the text of the argument.

By itself, this response can only be part of an account of our target reasoning, because it has not yet been shown that the constitutive inferences are valid—and, as far as I am aware, Kant nowhere addresses this question. But Kant seems to have a potentially attractive in-principle response to the Generality Objection. What is needed now is an assessment of Kant's notion of 'construction in pure intuition'; and of what he calls intuition's 'guiding role'.

Let us start with the difficult notion of construction in pure intuition. It may be helpful to review some of the terminology. First, then, we need to recall the fundamental distinction for Kant between intuitions and concepts. Any claim to knowledge requires representations of both types. Intuitions are singular and unmediated (re)presentations of objects given in sensibility.[103] The faculty of intuition is the capacity we have to form, and be receptive to information conveyed by, intuitions. Concepts, by contrast, are general representations whose source is the understanding.[104] Overall:

> [A cognition] is either an **intuition** or a **concept** (*intuitus vel conceptus*). The former is immediately related to the object and is singular, the latter is mediate, by means of a mark, which can be common to several things.[105]

---

[103] Kant also apparently believes that intuition can in principle be intellectual (cf. B308, also A256/B312); such would be the intuition of a divine being (B145). But we can ignore this point here.

[104] E.g. A19/B33; A66/B91.

[105] A320/B377.

There has been extensive debate as to how exactly the 'mediated/unmediated' distinction is to be understood, here and elsewhere. Rather than revisit this debate, I shall simply take it that Kant's point is that concepts represent by means of certain marks or features of the object that can in principle be common to other objects, while intuitions do not. Though intuitions may represent by means of marks, intuitive marks themselves have no generality as such.[106] It is important to bear in mind that singularity and generality for Kant are here properties of *types of representation*, not of objects represented. Kant takes concepts to be intrinsically general representations: for an object to fall under a concept is for it to belong to a kind of which there are, or might in principle be, other instances. Intuitions, however, are singular representations, as suggested by the *Jäsche Logic*:

> All modes of knowledge, that is all representations related to an object with consciousness are either intuitions or concepts. The intuition is a singular representation (*repraesentatio singularis*), the concept a general (*repraesentatio per notas communes*) or reflected representation (*repraesentatio discursiva*).[107]

Second, as singular representations of objects, intuitions make a distinctive contribution to cognition of objects; they present information about objects. But to understand an object as an object of a given type, also requires a concept. This seems to be the point of the following passage, again from the *Jäsche Logic*:

> When a savage sees a house in the distance, the use of which he does not know, he has the same object before him as another who knows it as a dwelling furnished for men. But as to form, this cognition of one and the same object is different in both. In the one it is mere intuition, in the other intuition and concept at the same time.[108]

---

[106] Here I am broadly following the discussion of marks in Smit 2000. The philosophical and exegetical issues are too complex to be helpfully summarized here, unfortunately; but they receive detailed analysis in the article above, which also cites much of the relevant recent literature.

[107] Kant 1974, p. 96.

[108] Kant 1974, p. 37f.

It seems Kant's thought here is that someone who lacks a given concept, and who is incapable of forming a thought or making a judgment employing that concept as a result, may nevertheless be sensitive to an object's being *this way* or *that way* in his visual experience. In such a case, 'this way' or 'that way' do not pick out features that the subject is generally aware of *as* such—it is not that being 'this way' is something that she could generalize to other objects, since this would make it conceptual—and possession of this sensitivity does not rely on any conceptual or propositional grasp of what it is for such an experience to be of that kind.

It follows that, strictly speaking, there cannot be such a thing for Kant as an intuition *that* P; this is a manner of speaking. For such a propositional grasp would contain a conceptual component that is not available to a reasoner who is merely having an intuition. And in saying that such a reasoner is having an intuition *of* an object, care must be taken not thereby to ascribe on the basis of intuition alone any awareness on the reasoner's part *that* the object in question has such and such a property or set of properties.

Third, intuitions can be either empirical or *a priori*. *A priori* or pure intuitions are intuitions that do not derive from sense experience, and that lack sensation as a result.[109] Kant analyses appearances (i.e. 'the undetermined object[s] of empirical intuition', A20/B34) in terms of 'form' and 'matter'. The 'matter' is that which is given in sensation, and is therefore *a posteriori*. The 'form' is the unifying structure sensations themselves must have if we are to be conscious of them as a single coherent experience, rather than as a mere buzz. According to Kant, in relation to external objects this structure is spatial, and given by pure intuition: it is only because we possess an *a priori* capacity to understand our outer sensations as spatially ordered that we can have experience of objects at all. Indeed, for Kant the science of geometry is only possible in virtue of pure intuition; and this is borne out for him by the fact that we can have knowledge *a priori* of the truths of geometry, without any epistemic reliance on experience or sense-perception.[110]

Fourth, it is intuition—and specifically empirical intuition—that is supposed to guarantee the objective meaning of a thought, by ensuring that the thought relates to an actual or possible object of knowledge. As Kant remarks:

---

[109] A20-2/B34-6. Note that in the discussion that follows, for the sake of convenience, I restrict attention to intuitions in 'outer sense'; that is, in relation to objects distinct from ourselves, as opposed to objects in 'inner sense' such as mental states.

[110] A24/B39; A47/B65ff.

> Now the object cannot be given to a concept otherwise than in intuition, and, even if a pure intuition is possible *a priori* prior to the object, then even this can acquire its object, thus its objective validity, only through empirical intuition, of which it is the mere form. Thus all concepts and with them all principles, however *a priori* they may be, are nevertheless related to empirical intuitions, i.e. to *data* for possible experience. *Without this they have no objective validity at all, but are rather a mere play, whether it be with representations of the imagination or of the understanding.* One need only take as an example the concepts of mathematics, and, first, indeed, in their pure intuitions. Space has three dimensions, between two points there can be only one line, etc. Although all these principles, and the representation of the object with which this science occupies itself, are generated in the mind completely *a priori*, they would still not signify anything at all if we could not always exhibit their significance in appearances (empirical objects). Hence it is also requisite **to make** an abstract concept **sensible**, i.e. display the object that corresponds to it in intuition, since without this the concept would remain (as one says) without **sense**, i.e., without significance.[111]

Note that the claim here is not (or not merely) that empirical intuition ensures the applicability of mathematics to the everyday world; it is that, according to Kant, empirical intuition is required for mathematical principles to have any 'objective validity' at all. This should not, I think, be taken to imply that mathematical principles have no meaning at all in the absence of empirical intuition. Kant acknowledges that purely conceptual thought can have meaning. But such thought can do no more than allow us to grasp the logical possibility of an object of experience: what Kant calls the 'formal conditions of an experience in general.'[112] Something more is required for objective validity:

> That in such a concept [i.e. a 'pure concept, which nevertheless belongs to experience'], no contradiction must be contained is, to be sure, a necessary logical condition; but it is far from sufficient for the objective reality of the concept, i.e. for the possibility of such an object as is thought through the concept. Thus in the concept of a figure that is enclosed between two straight lines there is no contradiction, for the concepts of two straight lines and their intersection contain no negation of a figure; rather, the impossibility rests not on the concept in itself, but on its construction in space, i.e. on the conditions of space and its determinations.[113]

[111] A239/B298. Emphasis added to the third sentence.

[112] A220-1/B267-8.

[113] Ibid. My italics.

That is to say, in effect, that without intuition we could not think of a figure as (or, better, as representing) an instance of the relevant concept(s). It is only once we can entertain such thoughts that the impossibility of a figure contained by two intersecting straight lines becomes manifest.

Thus Kant is appealing to a notion of objective or real significance given by a claim's applicability to objects of possible experience. Without empirical intuition, mathematical claims would have meaning, but there would be no guarantee of their real significance. This applies in principle to 'pure' as much as to 'applied' mathematics, though Kant does not use this modern distinction as such.

Fifth, the contrast between empirical and pure intuition is brought out by Kant's important distinction between the 'image' of a given geometrical concept, and the 'schema' of that concept; and it is this that Kant specifically invokes in order to avoid the Generality Objection mentioned earlier:

> Now this representation of a general procedure of the imagination for providing a concept with its image is what I call the schema of the concept. In fact it is not images of objects but schemata that ground our pure sensible concepts. No image of a triangle would ever be adequate to the concept of it. For it would not attain the generality of the concept, which makes this valid for all triangles, right or acute etc., but would always be limited to one part of this sphere. The schema of the triangle can never exist anywhere except in thought, and signifies a rule of the synthesis of the imagination with regard to pure shapes in space.[114]

I think Michael Friedman must be right to interpret Kant's notion of an image as referring, in the context of geometry, to any particular diagram produced by a given construction procedure, and a schema as a representation of the construction procedure itself. 'The rule of the synthesis of the imagination' mentioned here is then further described in the Axioms of Intuition:

> If I say: 'With three lines, two of which taken together are greater than the third, a triangle can be drawn,' then I have here the mere function of the productive imagination, which draws the lines greater or smaller, thus allowing them to abut at any arbitrary angle.[115]

[114] A141/B180.

[115] A164/B205; cf. Friedman 1992, p. 124.

We are now in a somewhat better position to assess Kant's notion of construction in pure intuition. As I interpret it, Kant's point is this: when a reasoner draws or visualizes a diagram of a triangle, she is not merely using concepts acquired as a result of grasping Euclid's definition. Rather, she is relying upon a (for Kant, non-conceptual) sensation-independent capacity to represent something as occupying (actual or visualized) space. Though diagrams of triangles can be perceived in empirical intuition, Kant takes it that a reasoner can have an image of a geometrical triangle via the exercise of her visual imagination that is free of sensory input, and this would be pure, or *a priori*. Moreover, it is the latter that is epistemically primary, if the faculty of intuition is to remain independent of the senses.

The source of the generality of Euclid's conclusion is then this, according to Kant; that a reasoner who can grasp the relevant construction procedure in a given argument has thereby grasped a rule (a 'schema') by which any object of the requisite kind can in principle be visualized or drawn. Thus for Kant, the justification provided by Prop. I.32 does not derive from mere generalization of the visual awareness of a particular diagram or image. It is the knowledge of which figures can and cannot be constructed from a given sequence of construction procedures that constrains the generality and applicability of a given geometrical property, and so provides the warrant for claims about such properties.

All well and good. But though Kant is clearly on to something here, it should be evident that this more developed account is highly problematic. I will mention three specific worries. The first concerns the extreme lack of clarity as to what is meant by a schema. Recall that on Kant's official position, there are just two kinds of representation: intuitions and concepts. But schemata are neither intuitions nor concepts; nor can they be if they are to serve the mediating function Kant gives them between intuitions and concepts. So are they a third kind of representation?

The second worry is that this reconstruction is unsatisfactorily dependent on Kant's overall account of pure intuition. In the absence of a showing that space is a pure form of intuition, we will need independent reason to believe that pure intuition is a prerequisite to the ability to represent objects as in space at all; and more specifically, it will need to be shown that there can be a pure visual image of a triangle—i.e. a visual representation that is entirely free of all sensory content—as Kant seems to claim.

The third worry is perhaps the most damaging. Recall that Kant's position is, as noted, that the representational scope of the diagram in Euclid's argument includes all and only the figures that can be constructed by following the relevant construction procedure. Given that a schema is supposed to be a representation of a rule governing the exercise of pure imagination, this seems to imply that the reasoner determines the generality of the diagram by visualizing and then indefinitely iterating different alternative constructions 'in her mind's eye'. But I argue in Postscript 2 below that indefinite iteration is not a route to justified belief in Euclid's conclusion, and that in any case this is not Kant's view. Similar considerations rule indefinite iteration out as an explanation of the reasoner's grasp of the generality of the diagram. So if we are to respect Kant's idea that the relevant construction procedure determines the representational content of the diagram, this claim needs to be understood in a different way; and I shall argue that we can so understand it, and also identify a broadly Kantian solution to the Generality Objection, in Chapter 10.

7.5 Intuitive Guidance

So far much of the discussion has focused on Kant's view as it bears on the interrelated issues of the representational scope of the diagram and the generality of the justification offered by Euclid's argument. These topics have received some attention in the Kant literature.[116] But there is an important further aspect, which bears on the general nature of the inferences relating to the diagram, and this has received much less attention from Kant's commentators to date; perhaps from a faulty background assumption that these inferences are not, or cannot be, valid or knowledge-yielding.[117]

We can approach this topic by asking what Kant means by his insistence in the Doctrine of Method that the geometer follows Euclid's argument by reasoning 'through a chain of inferences that is always guided by intuition'.[118] One way to understand the notion of 'guidance' here might be in terms of a justificatory appeal by 'reading off' properties from the diagram. Michael Potter comes close to this reading, in suggesting that Kant believed that what underwrites the existence of a constructed point C in the famous equilateral triangle of Prop. I.1 is the fact that the lines of the diagram or imagined figure actually cross.[119] But if this is just Kant's view, then it is in serious difficulty from the outset, since we have already noted that the justificatory appeal to the figure by itself is fallacious. On the other hand, if 'guidance' is to be taken purely psychologically and as having no

---

[116] Notably in Friedman 1992; but see also Howell 1973, Parsons 1983 Ch. 5 and Postscript, and Smit 2000. E.B. Williams has rightly stressed the centrality of intuitive guidance to me in conversation.

[117] An assumption typically made by exponents of the 'logical' interpretation of Kant's philosophy of geometry. I examine some aspects of this interpretation further in Postscript 1.

[118] A717/B745.

[119] Potter 2000, p. 47.

epistemic significance, then it is not clear why we should think of the diagram as contributing to justification, according to Kant.

Can we do better than this? Kant seems to offer as part of his explanation of the role of intuition a contrast between the type of construction to be found in algebra, and that to be found in geometry:

> But mathematics does not merely construct magnitudes (*quanta*), as in geometry, but also mere magnitude (*quantitatem*), as in algebra, where it abstracts entirely from the constitution of the object that is to be thought on accordance with such a concept of magnitude. In this case it chooses a certain notation for all construction of magnitudes in general (numbers), as well as addition, subtraction, extraction of roots etc., and, after it has also designated the general concept of quantities in accordance with their different relations, it then exhibits all the procedures through which magnitude is generated and altered in accordance with certain rules in intuition; where one magnitude is to be divided by another, it places their symbols together in accordance with the form of notation for division, and thereby achieves by a symbolic construction equally well what geometry does by means of an ostensive or geometrical construction (of the objects themselves), which discursive cognition could never achieve by means of mere concepts.[120]

Knowledge of algebra and geometry alike requires intuition for Kant, for both algebraic and geometrical reasoning rest on constructions in pure intuition. But the construction procedures are, he claims, of different kinds: in the case of geometry the procedure is 'ostensive', and the construction is 'of the objects themselves'; while in the case of algebra the procedure is symbolic and 'abstracts from the constitution of' the object(s) represented.

I interpret Kant's point here to relate to two different ways in which construction procedures can be used to represent mathematical objects or functions. Symbolic representation uses symbols such as numerals and function signs; and it is indirect, in that the reasoner knows what a certain symbol represents only in virtue of background knowledge as to what assumptions link it with its object or target of representation. These assumptions are conventional in nature: any symbol can in principle, given a suitable set of background assumptions, be taken to represent any object. Thus, for example, all of 'x', '*' and '.' are standardly used to represent multiplication, but so for example could ' ', ' ', or '⌘', if suitable conventions could be established. There is no in-principle constraint (though there may be constraints of ease of use etc.) on the actual form of representation or notation to be adopted. So symbolic representation permits a choice of no-

120 Ibid.; emphasis added in the last sentence.

tation between alternatives, all of which can bear the requisite representational relation to what they represent; and there are different notations available for numbers and arithmetical operations, for example, as Kant would have been aware.[121]

Ostensive representation, by contrast, is *direct*: the diagram must be such that it can itself reliably be recognized as a representation of an instance of the intended category. This is an important in-principle constraint on ostensive representation, which has the effect of restricting the availability of alternative representational forms; there will be few if any alternatives to a diagram or figure of a triangle as a means to represent a geometrical triangle ostensively. Contextual assumptions notwithstanding, ostensive representation is not a merely conventional relation between diagram and object(s).

Euclid's geometry, as given in what I have termed the Euclidean Presentation, does not use purely ostensive representational forms. First, it uses letters as labels, and these are symbolic, not ostensive. Secondly, words of natural language (as in the sentences of the Propositions) are generally symbolic in the sense above. Euclid's geometry is, then, strictly a mixed or heterogeneous system, in that it uses both ostensive and symbolic representations, the latter including the words of the text. But the use of diagrams gives it a heavily ostensive character.

We can now see what Kant appears to have in mind in referring to ostensive constructions as constructing 'the objects themselves'. An initial instruction bids the reasoner draw a diagram of the relevant object(s): in Prop. I.32, this is in the opening sentence of the Setting-Out ('Let ABC be a triangle'). Every construction procedure used to draw a given diagram can in principle be used to depict each of the objects represented by the diagram. There is thus, given the relevant background assumptions, a correspondence between the elements and properties of the diagram and those of its target or object(s): a correspondence that preserves in the diagram what we should now term the mathematical structure of the object(s). The reason why Kant claims that ostensive presentation is of 'the objects themselves' is that this structure preserves all the relevant geometrical properties of the class of geometrical objects in question. Thus in looking at the diagram it is in each case 'as if' a reasoner is looking at a geometrical object (or spatial configuration of geometrical objects). Such reasoning can be general in that it relates to a class of objects all of which conform to conditions of construction, and thus possess the relevant structure. So any claim made by

---

[121] I have one serious reservation as to Kant's account of symbolic construction, but this does not bear on the distinction as such, and need not concern us here. See Postscript 3 below.

reasoning with the diagram will apply, *mutatis mutandis*, to any member of that class.

What, then, is the force of Kant's insistence on the 'guiding role' of intuition in geometrical reasoning? Should we understand this guidance psychologically, as referring to the role of the diagram in prompting inferences, or epistemically, as referring to the role of the diagram in contributing to justification?

I suggest that Kant has *both* of these roles in mind. Psychologically, he reminds us that on drawing the auxiliary line the geometer 'sees that here there arises' an angle adjacent to an existing angle—and it is by seeing the alternate and opposite angles on the diagram that the geometer is prompted to apply the relevant rules (e.g. Prop. I.29). Epistemically, Kant stresses that it is intuition, via ostensive construction, that permits the reasoner to gain an understanding of the diagram on which the diagram can serve to justify claims about geometrical objects. For the diagram conveys structural information about the objects it represents, when understood in accordance with appropriate background conventions, and this feature is exploited by the reasoner in making inferences with the diagram.

On this reading, Kant takes the reasoning here—including the reasoning with the diagram—to be valid reasoning. Euclid's argument is then 'illuminating' because the diagram is used ostensively, to represent geometrical 'objects themselves'; and it is valid 'universally' because a reasoner's grasp of construction procedures underwrites the argument's general conclusion. We can now better appreciate, for Euclid's geometry at least, why Kant claims that it is construction—and in particular ostensive construction—that generates the 'apodeictic' certainty of mathematics.[122] For it is in ostensive construction, and in the capacities that underlie it, that Kant locates the source of the illumination, directness and generality of this kind of reasoning.

But we should again note that there are two critical gaps in Kant's exposition: first, he needs to show, not merely that a diagram can preserve structural information about its objects as described, but that the inferences with the diagram in Euclid's argument are valid inferences; second, he needs to show how a reasoner can be justified, indeed justified *a priori*, in believing the angle sum claim in full generality. I return to these questions in Chapter 10.

[122] A713/B741.

7.6 Summary

This chapter has argued that we can identify a Kantian View of the reasoning involved in following Euclid's argument. Kant himself held the Kantian View, and his treatment of geometry in the Doctrine of Method has been discussed in detail. According to the interpretation developed here, Kant is using a case study of geometrical reasoning to illustrate his general claim that geometry (and ultimately, of course, mathematics generally) is synthetic *a priori*; and the view of reasoning that this implies can be discussed in isolation from Kant's wider claims for Euclid's geometry, or for intuition.

What emerges is, in effect, a theory of a certain kind of visual thinking or reasoning. On this theory, in following Euclid's argument, a reasoner can use a figure or diagram to represent geometrical objects of a given type *a priori*; the diagram can permissibly be altered in accordance with certain given construction procedures; such alterations to the diagram preserve structural information about the object(s) represented; the generality of the construction procedure is sufficient to justify the generality of the conclusion of the argument; and a reasoner who can appropriately grasp the relevant construction procedure, and who can follow the argument in relation to the diagram, can thereby be justified in forming a general belief *a priori* in the truth of the conclusion. Moreover, we can use Kant's distinction between ostensive and symbolic construction, within the context of the reconstructed theory, to give a plausible in-principle explanation of Kant's claim that intuition offers 'guidance' and 'illumination'.

7.7 Postscript 1: Kant and the 'Logical Interpretation'

The interpretation advanced here treats the Doctrine of Method seriously: as, in effect, advancing a coherent and in many respects rather persuasive claim for the epistemic value of a particular kind of reasoning. By contrast, the consensus amongst Kant's commentators has long been that Kant's claims in the Doctrine of Method are obviously mistaken. This view was famously expressed by Bertrand Russell, as noted in Chapter 1:

> Kant, having observed that the geometers of his day could not prove their theorems by unaided arguments, but required an appeal to the figure, invented a theory of mathematical reasoning according to which the infer-

> ence is never strictly logical, but always requires the support of what is called intuition.[123]

Though this view has been extremely influential among philosophers of many different stripes, it has been developed in recent years into what has been termed the 'logical interpretation' of Kant's philosophy of geometry. On the logical interpretation, the function of intuition for Kant here is, in effect, to compensate for deficiencies in the then-available logic. In this Postscript, I want to contrast this interpretation with the view developed above.

The logical interpretation has been developed most explicitly in the work of Michael Friedman.[124] Friedman claims, of Kant's discussion of the Euclidean Presentation:

> In contending that construction in pure intuition is essential to this proof, Kant is making two claims that strike us as quite outlandish today. First, he is claiming that (an idealized version) of the figure we have drawn is necessary to the proof. The lines AB, BC, CE, and so on are indispensable constituents; without them the proof simply could not proceed. So geometrical proofs are themselves spatial objects. Second, it is equally important to Kant that the lines in question are actually drawn or continuously generated, as it were. Proofs are not only spatial objects, they are spatio-temporal objects as well...
>
> Kant's conception of geometrical proof is of course anathema to us. Spatial figures, however produced, are not essential constituents of proofs, but, at best, aids (and very possibly misleading ones) to the intuitive comprehension of proofs. Whatever the intended interpretation of the axioms or premises of a geometrical proof may be, the proof itself is purely 'formal' or 'conceptual' object; ideally, a string of expressions in a given formal language.[125]

Although Friedman regards the claims in the Doctrine of Method as mistaken, his overall approach is sympathetic to Kant. It recalls that, on a standard view, the reason for many if not all of the logical gaps in Euclid's geometry is that it lacks existence axioms, a theory of order governing points in the line, and modern concepts of continuity, denseness etc. On Friedman's view, Kant is implicitly aware that he has no means in the

---

[123] Russell 1919, p. 145.

[124] See especially Friedman 1992, Chapters 1 and 2; and Friedman 2000.

[125] Friedman 1992, pp. 57-58. I am assuming here that Friedman is speaking 'in his own voice'; that is, that he holds the views described here.

(canonically syllogistic, subject-predicate) logic then available to express certain desired concepts of continuity, infinity and infinite divisibility. The function of intuition is, in this regard, to permit such concepts to be represented. A concept of continuity is, the suggestion goes, given representation for Euclid by the motion of a mathematical point (an idealized stylus) drawing a line; a concept of infinity is given representation by the reasoner's (idealized) ability to iterate, for example, the production of a line segment an indefinite number of times by application of Postulate 3; and a concept of infinite divisibility is given representation by the (idealized) iterable bisection of a line segment according to Prop. I.10. These actions all are, or involve, implementations of construction procedures. Moreover, construction procedures are required, according to Friedman, if diagrams in Euclid are to give rigorous representation of the relevant geometrical concepts. Thus Friedman says of Euclid's construction of a circle in Postulate 3:

> The underlying idea [behind Kant's claim that this construction 'first generates the concept of a figure' at A234/B287] ... is that the existential proposition corresponding to this construction—that for any point and any line there is a circle with the given point as centre and the given line as radius—cannot be conceptually expressed for Kant. In mere syllogistic logic this existential proposition cannot, strictly speaking, even be stated (as we would now put it, it involves the form of quantificational dependence $\forall\forall\exists$). The only way even to think or represent this proposition—so as, in particular, to engage in rigorous geometrical reasoning thereby—is by means of the construction itself.[126]

Thus on this view Kant believes that it is the possibility that the Euclidean Presentation provides to construct diagrams on paper or figures in imagination that allows representation of these foundational geometrical concepts, and other concepts from complex diagrams constructible from the basic ones; and it is by means of intuition that a reasoner can come to possess these concepts.

On this interpretation, then, the role of intuition for Kant is primarily as a substitute for what would in modern terms be logical forms of representation using quantifiers.[127] The stage is now set for Friedman to claim that this view has a crucial drawback, however; one that we can appreciate but Kant could not. This comes in Kant's mistaken belief that construction procedures are justificationally required to secure the relevant order proper-

[126] Friedman 1992, p. 126.

[127] See, for example, Friedman 1992 pp. 70-71.

ties of the geometrical line; mistaken, since these properties can only be rigorously formalized, as discussed, via a modern theory of order in the style of Hilbert. We can know, thus, as Kant could not, that the diagrams in Euclid are 'inessential constituents of proofs, but at best aids... to the intuitive comprehension of proofs', which are purely formal objects.

Friedman's interpretation of Kant is subtle and illuminating; and I have only been able to sketch a small part of the overall line of thought. Nevertheless, I think this much is enough to suggest that it—and the logical interpretation generally—faces a number of difficulties as an account of Kant's thinking here, on both internal and external grounds.

The first difficulty relates to a question of exegesis. Friedman introduces his discussion via a consideration of the Doctrine of Method; he uses this to raise modern worries about the order properties of the line in Euclid, and then seeks to show that these in turn generate questions as to the representation of continuity and infinity for Kant, to which the latter is alive (although perhaps only implicitly) in the Aesthetic and Analytic. But these logical issues, though important, are plainly not ones that Kant has principally in mind in the Doctrine of Method, and do not help, at least in any direct way, to explain his thought there.

By contrast, the Doctrine of Method passage is much more clearly focused on the *epistemology* of the different processes of reasoning themselves. Kant is asking: given that there are these two apparently different processes of reasoning—mathematical and philosophical—how do they differ, and how does each justify and guide us in coming to certain conclusions? What seems to distinguish the geometer's reasoning is, at least in the first place, that it relates to a particular diagram, has a visual component, and yields 'evident and universally valid' conclusions. It is these phenomena that require explanation; and the putative role of intuition in guaranteeing representation of certain concepts of continuity etc., while it may be relevant overall, is not the target here. Indeed, as its opening words make clear, the wider point of the Doctrine of Method as a whole is, precisely, to provide the plan and methods by which the various constituents (the 'elements' of pure reason) can be combined into an entire system of pure speculative knowledge. How they are to do so centrally concerns the epistemology of the methods of reasoning involved.

The second difficulty may seem rather fussy; but it has a wider point. Recall that Friedman interprets the Doctrine of Method passage as making two apparently 'outlandish' claims. The first of these is that '(an idealized version) of the figure we have drawn is necessary to the proof. The lines AB, BC, CE, and so on are indispensable constituents; without them the proof simply could not proceed. So geometrical proofs are themselves spatial objects.' But this is surely not Kant's claim. Rather, his claim relates to the proof (strictly speaking: argument) *as presented by Euclid*; he is claiming that the figure is necessary to the Euclidean Presentation of the argument, that the presentation could not be understood without it. Far from being outlandish, this claim seems to be true, for reasons discussed in Chapter 6. Does it mean that the presentation must be spatial? If the presentation takes a written form and contains a diagram, then it will occupy physical space. If the presentation is simply thought through in the mind, then it will not occupy physical space. But even in this case it is nevertheless plausible that the figure as visualized presents spatial, specifically two-dimensional, information. So Kant's point here is—in line with the focus of the Doctrine of Method on the epistemology of reasoning—to focus on the nature of our interaction with a specific presentation of the argument. That is, Kant is seeking an explanation that conforms to what he takes to be the facts, and specifically the phenomenology, of the way in which the reasoner can correctly follow Euclid's argument.

Finally, and most crucially, it remains quite unclear why intuition is supposedly required to compensate for deficiencies in logic. Why, for example, should the absence of a formal language of quantifier logic within which to represent the existential proposition corresponding to Postulate 3 have any bearing on the ability of Kant—or a geometrical reasoner—to give that proposition conceptual expression? Why, for example, could it not be sufficient for this purpose simply to express it with the sentence 'For any point and any line there is a circle with the given point as centre and the given line as radius'? It is not clear that Kant is moved at all by the expressive limitations of logic as he then knew it. Quite the contrary: if the reading of Kant's distinction between ostensive and symbolic construction given here is correct, a sentence such as that above—or a sentence in a formal language of quantifier logic—would be deemed symbolic, not ostensive, precisely because it did not present, and preserve through a given process of reasoning, what we might term the spatial content of the intuition of the diagram. Moreover, if we regard intuition as a mere substitute for logical representation, the main thrust of Kant's insistence on the guiding role of intuition in geometrical reasoning, which allows the reasoner to construct and reason as to 'the objects themselves', is lost.

Notice that I am not here disputing the positive claim that intuition may in principle be invoked to play a quasi-logical role for Kant in the way Friedman describes. However, on the interpretation I propose Kant's concerns are not purely logical, and the role of the diagram is not merely to compensate for deficiencies in his logic. Rather, we need to take what he says in the Doctrine of Method at face value. If we do so, we can see that Kant is here primarily seeking to explain, not the logical presuppositions of a Euclidean argument, but the distinctive patterns of reasoning that occur in relation to the diagrams within the Euclidean Presentation. If this is so, then the role of intuition in reasoning is distinguishable from any role it may have in, for example, guaranteeing the order properties of the line; one could, as it were, add Hilbert's theory of order (in a suitable form) explicitly as background assumptions to the Euclidean Presentation and still reason wholly or partly in the way Kant has in mind in the Doctrine of Method.

7.8 Postscript 2: Iteration and Justification

In this Postscript I want to consider an alternative reconstruction of Kant's view of the apparent justification offered by following Prop. I.32. Ian Mueller remarks in a very similar context:

> It is natural to ask about the legitimacy of such a proof ... How can one move from an argument based upon a particular example to a general conclusion, from an argument about a straight line AB to a conclusion about any straight line? I do not believe that the Greeks ever answered this question satisfactorily ... Of course, insisting that the particular argument is sufficient to establish the general *protasis* is not a justification, but it does amount to laying down a rule of mathematical proof: to prove a particular case is to count as proving a general proposition.[128]

If Mueller is right here, the generalization in Euclid's argument is unwarranted, since we have no reason to accept the rule of mathematical proof he proposes; different types of triangle can have different geometrical properties.

Are we compelled to accept Mueller's conclusion? Kant argued for the opposite view, as we have seen. But it is sometimes suggested that Kant's account of intuition has an important weakness at precisely this point. Recall that an intuition for Kant is an immediate singular (re)presentation. A common approach here is to try to justify the generalization in terms of an indefinitely long process of iteration or repetition; for example, this approach is adopted by Reviel Netz in his recent book on deductive reasoning in Greek geometry.[129] With this in mind, the interpretation might be: that according to Kant the reasoner takes Euclid's argument to go through for a single (intuitively presented) case, which intuition might be of an imagined

---

[128] Mueller 1981, p. 13.

[129] Netz 1999.

figure, and so be *a priori* or 'pure'; and that the reasoner reaches the general conclusion by iterating the result of this individual case indefinitely. But then, the objection goes, this is effectively an argument with an infinite premiss.[130] It is of the following form:

Fa Fb Fc ...

. . .

. . .

. . .

Ga Gb Gc

$\forall x\,(Fx \rightarrow Gx)$

But such arguments, the objection goes, cannot be followed by finite minds. So they cannot justify their conclusions.

There is an important philosophical question here as to whether and how we can grasp infinite rules of this kind, and someone prepared to embrace this alternative might accept this as a reconstruction of Kant but dispute whether it is more than an embarrassment. But in fact I suggest that this would not be a correct reading of Kant at all. I argued above that Kant's account is designed to be psychologically plausible. If we accept this, then it makes this interpretation problematic from the outset, since it is quite evident that the reasoner does not even attempt to follow any indefinitely long rule, nor is there any apparent feeling of obligation to do so. Secondly, note that Kant does not mention (anything amounting to) iteration, nor describe any iterative process, in the Doctrine of Method. What he actually says is that the 'individual drawn figure ... serves to express the concept without damage to its universality.'[131] Thirdly, we cannot properly motivate the objection by appealing to the singularity of intuition, for the type of representation that underwrites the generalization for Kant is not an (either empirical or pure) intuition, but the schema of a concept. That is, it is not the case that the singularity of intuition somehow forces us to understand the generalization as involving repetition of the argument (or diagram) for different geometrical cases. Fourthly, iteration is at odds with the overall thrust of Kant's account, since it specifically raises the possibility that the justification is empirical, by induction on visualized figures. This would make it similar to the positions explored and rejected in Chapters 4 and 5. As I noted there, if there is iteration in following Euclid's argument,

---

[130] On this see, for example, Potter 2000, p. 47.

[131] A714/B742.

its role is not justificatory; it may only be to fix the concept in the reader's mind.

For these reasons, then, we should reject the interpretation offered above as a reading of Kant.

7.9 Postscript 3: Symbolic Construction in Intuition

The goal of this Postscript is to raise a serious question about Kant's account of symbolic construction. Recall that on Kant's official definition, intuitions are immediate singular representations. Thus an empirical intuition of a diagram is an immediate singular representation of a diagram, an empirical intuition of a numeral is an immediate singular representation of a numeral; and the same goes in principle for function signs in arithmetic, algebraic symbols etc. Ostensive construction is direct in that, when executed and understood correctly, it preserves certain structural information as between the constructed diagram and the object(s) it represents. It is thus fairly evident why Kant insists that ostensive construction is intuitive, for it preserves the direct contribution to cognition made by the diagram as a representation of its objects.

But why does Kant insist that symbolic construction is intuitive? After all, symbolic construction is not by this criterion immediate: the representational link between a symbol and its object(s) is an indirect one, and imposes no in-principle constraint on different potential forms of representation, as we have seen. A numeral (or function sign etc.) normally denotes only in virtue of a conventionally agreed relation with its object. It is quite unclear how someone could, even in principle, become aware of this relation without already possessing at least some of the relevant concepts. But this calls into question how, for example, intuitions of numerals as representations of numbers can be genuine intuitions at all. Of course, there is a spatial component to the construction of written arrays of symbols, and this may be what Kant has in mind. But even if one may have intuitions of numerals, it is only in virtue of the relevant mediating conventions that such intuitions can convey information as to—and so give cognitive access to—their objects, and this is surely conceptual. This point is independent of the question whether Kant has a plausible account of intuition of numbers via processes of reasoning that do not involve numerals and other symbols; he may do. In short: though the distinction between ostensive and symbolic construction is a valuable one, Kant's claim that symbolic construction is in his terms intuitive is, I suggest, highly questionable.

8

Making Room for a Neo-Kantian View

8.1 Introduction

The previous chapter completed our survey of the possible types of explanation of the reasoning involved in following Euclid's argument, according to the Framework of Alternatives set out in Chapter 3. If the argument so far is correct, then the Kantian View not only avoids the difficulties facing its competitor views, but also advances a promising though partial explanation of this visual thinking. However, it does so within a set of assumptions that many philosophers have found it hard to accept, and in the face of important further objections in principle. Can a similar view be persuasive that does not require these assumptions? In particular, can it meet these further objections? On both counts, I shall argue that it can.

The present chapter is intended to make room for a positive explication and defense of one version of the Kantian View: a neo-Kantian view, if you like. This name is meant in a literal and neutral way: not as implicitly affiliating the view in question with any neo-Kantian school, thinker or doctrine, nor as suggesting that there may not be more than one such view plausibly available. The neo-Kantian view advanced here (henceforth for convenience, the neo-Kantian view) is a version of the Kantian View because it accepts the distinctive claims of the latter: that the reasoning is *a priori*, and that the diagram makes a justificatory contribution. It is *neo*-Kantian because it does not follow Kant in many of the further claims he makes for his overall theory of intuition; it is, in effect, a different subspecies of the Kantian View. On several of the relevant issues, the neo-Kantian view simply does not take a position; on others, the position adopted is not Kant's.

Accordingly, this chapter describes the neo-Kantian view, and locates the main points at which it diverges from Kant's own views. It then assesses three further possible in-principle objections, over and above the Generality Objection already discussed in Chapter 7. These objections have been advanced in various forms by a number of philosophers, but I shall approach them as presented by Philip Kitcher, in a cumulative critique of Kant that forms part of his own case for an empiricist account of mathematical knowledge.[132] In Kitcher's writings, they constitute a convenient, accessible and influential body of criticism. I shall not offer any detailed defense of Kant's own position against these arguments, though I shall sketch possible lines of defense in places, and especially in a Postscript (Section 8.7). The question will be, rather, whether the neo-Kantian position advanced here is undermined by them, or by analogues of them; and if so, how it may appropriately be defended. There may of course be other worries, both to the overall project here and to its detailed implementation, but this discussion is designed to address at least these main in-principle lines of objection.

8.2 Kant vs. the Neo-Kantian View

With this in mind, it will be helpful to make clear from the outset the principal points at which Kant's own doctrines and the neo-Kantian view diverge. We can class the differences into three:

1. *The status of Euclid's geometry.* Kant took Euclid's geometry to be the science of space: an exact and necessarily true description of the spatial structure of the physical universe. The neo-Kantian view rejects this claim. On the neo-Kantian view, Euclid's geometry is best understood as a piece of pure mathematics. It is a further question, one to be settled by empirical investigation, whether or not the spatial structure of the physical universe is in fact Euclidean. It is possible to reason about a kind of mathematical 'space' using Euclid's geometry. When someone does this, she is reasoning not about the actual properties of physical space, but about what features or properties physical space would have if it were as described in that geometry.
2. *Intuition as a faculty of mind.* As described briefly in the last chapter, Kant postulates intuition as a fundamental mental faculty, and appeals to the workings of this faculty to explain, among other things, the synthetic *a priori* status of mathematical knowledge.

[132] Kitcher 1975, Kitcher 1984, Kitcher 2000. Kitcher 1984 recapitulates arguments against Kant originally advanced in Kitcher 1975, and it will be convenient to treat them together.

It is notable that Kant's is not a 'black box' appeal to intuition, on which little further is said, so that intuition comes to seem occult and non-explanatory. On the contrary, Kant has a great deal to say about his own specific notion of intuition, as we have seen, and it is this positive theory, rather than the fact of any missing explanation, that many philosophers have found objectionable. The neo-Kantian view avoids both these alternatives, since it does not postulate or invoke the workings of any mental faculty of intuition. On the neo-Kantian view, to put the matter very roughly, the key functions accorded by Kant to intuition as a faculty—the capacity to represent spatial content and to generalize—are carried out by means of the reasoner's possession of a certain conceptual repertoire, in conjunction with her visual imagination. The reasoner is able to form a geometrical concept of a certain kind; having done so, she can reason with the diagram by taking it to represent one or more instances of the concept. Finally, she employs a conceptual, and not intuitive, capacity to generalize.

Note that, though it rejects the claim that we have some faculty of intuition, the neo-Kantian view need not reject the claim that there is something 'intuitive' about the reasoner's geometrical concepts. On the view to be advanced here, the reasoner can form a geometrical concept by reflecting on her perceptual experience, and the concept so formed is spatial in that it enables its possessor to recognize, use and manipulate two-dimensional figures or diagrams as representing instances of the concept. This topic is discussed in detail in Chapter 10.

3. *Apriority*. Kant held that all and only necessary truths are knowable *a priori*, that *a priori* knowledge is certain, that *a priori* justification is indefeasible, and that truths knowable *a priori* are not knowable *a posteriori*.

   The neo-Kantian view is, of course, committed to the claims that there are mathematical truths knowable *a priori*, and that a reasoner may come to know at least one such truth by following Euclid's argument. But beyond that it need not make any further commitments. It is thus compatible in principle with views that claim that some contingent truths may be known *a priori*, that some necessary truths may be known *a posteriori*, that *a priori* knowledge may be non-certain, that *a priori* justification may be defeasible and that some truths may be known both *a priori* and *a posteriori*.

8.3 The Irrelevance Objection

With this in mind, I now turn to the various objections advanced by Kitcher against Kant. The first is what he calls the Irrelevance Objection, and this follows directly on from the Generality Objection, a version of which was considered in the last chapter. The Irrelevance Objection goes as follows:

> Kant believes we can gain *a priori* knowledge about the general properties of triangles by drawing and inspecting a particular triangle. But how do we come to generalize over the right properties and avoid generalizing over the wrong ones? ... Were Kant to suggest that we should only generalize over those properties which are determined by the *concept* of the triangle, *the process of constructing mental diagrams would simply be a vehicle for disclosing conceptual relations and Kant's position would become a conceptualist version of apriorism.* Thus Kant must conclude that the presented triangle has three types of property: those properties determined by the concept of triangle; those properties which reflect the structure which we necessarily impose on experience, and those properties which result from accidental decisions made in the construction. For his account to succeed we need a method of discriminating properties of the two latter types, so that we can legitimately generalize over the former and avoid generalizing over the latter. But to be able to do this is to have precisely that knowledge of the structure of experience for which Kant is attempting to account![133]

Thus the structure of the objection is to present a dilemma: either the generalization is conceptual or it is not. If it is conceptual, then it does not epistemically depend on intuition (and so Kant must, it seems, reject this alternative). But if it is not conceptual, then the figure or diagram must have, as well as conceptual and accidental properties, intuitive or mind-imposed properties; and there is then a question as to how the intuitive properties may be non-circularly differentiated from the accidental properties.[134]

How damaging is this worry to the neo-Kantian view? The view is committed to the claim that the reasoner is able to generalize in virtue of the exercise of a conceptual capacity. But this need not commit it to the claim, as the italicized clause above implies, that the process of constructing the diagram is 'simply' a vehicle for disclosing conceptual relations. What Kitcher means by this is that the explanation in terms of the diagram becomes redundant if it invokes a purely conceptual capacity to generalize:

---

[133] Kitcher 1984, p. 51, emphasis added.

[134] Coffa 1991 p. 46 seems to express a similar concern.

> [Kant's solution in terms of schemata] seems to make the exhibition of a particular triangle in intuition quite unnecessary. For if all that we are allowed to do is to draw out features of triangles prescribed by the schema of the concept 'triangle', then we can do this by conceptual analysis alone.[135]

This statement is inaccurate as a summary of Kant's notion of a 'schema' and its role in this kind of reasoning, which we discussed in the previous chapter. A schema or representation of a construction procedure is, on Kant's account, what determines the representational scope or generality of the diagram; but, as I have interpreted Kant (see Section 7.5 above), the diagram itself—and not the schema—is used to justify certain claims in Euclid's argument. Appeal to the nature of schemata as such gives no reason to think that the diagram itself is unnecessary here, as Kitcher claims.

But the deeper question is: 'unnecessary' to what? As noted in Section 7.7, in line with the overall discussion in the Doctrine of Method, Kant's focus here is not on geometrical knowledge as such, but on the epistemology of the reasoning required to follow a given argument in Euclid. The same is true for the neo-Kantian. It is the apparent epistemic necessity of the diagram to *this kind of reasoning*, and not to geometrical knowledge *per se*, that is the target of explanation here. Hence Kitcher's objection is off-target. No consideration has yet been advanced against the former view.

It would be hard to argue that diagrams are necessary for any process of reasoning by which someone might acquire geometrical knowledge. But, for the reasons described in Chapter 6, it is plausible that a diagram (or figure) is epistemically necessary to the kind of reasoning required to follow Euclid's argument. The neo-Kantian view can grant that there may be a valid counterpart means to reason to the angle-sum claim (or a similar claim) that is non-diagrammatic, while denying that this has any bearing on the question at issue. It can, then, safely accept the first horn of the dilemma posed by the Relevance Objection; it can concur that the reasoning here is conceptual, without conceding that the diagram or figure is epistemically unnecessary to the process of reasoning itself, or that there is nothing at all recognizably 'intuitive' about such a process.[136]

Note that Kitcher does not argue for his claim that 'if all that we are allowed to do is to draw out features of triangles prescribed by the schema of the concept "triangle", then we can do this by conceptual analysis alone.' Rather, he simply appears to assume that it is somehow a fact about concep-

---

[135] Kitcher 1975, p. 43.

[136] Does Kant's position have the resources to meet the Irrelevance Objection? I argue in the Postscript to this chapter that it does.

tual analysis that a distinctive kind of reasoning that employs it can never be epistemically reliant on diagrams or figures. This assumption is, I have argued, mistaken. But since the assumption is itself symptomatic of deeper and more widely held preconceptions as to the epistemic status of diagrams, a brief consideration of why Kitcher appears to hold it may be of interest. The key point, I suggest, is that Kitcher generally associates concepts with linguistic meanings; indeed, he devotes the chapter on 'Conceptualism' in Kitcher 1984 to a detailed exploration of the relation between linguistic abilities and conceptual knowledge, defending this connection against Quinean-type criticisms and ultimately locating it within his preferred psychologistic and empiricist account of mathematical knowledge. By 'conceptual analysis', then, Kitcher appears to have linguistic analysis canonically in mind.

The first thing to note here is that this is far removed from Kant's picture, on which the diagrams in Euclid are 'ostensive' representations, and the relevant thoughts depend on intuition for the representation of spatial content. Understanding concepts in terms of linguistic meanings, by contrast, predisposes analysis in a quite different direction. Take a counterpart argument in natural language to the angle-sum claim. Here, as in a formal logical language, the argument is—in the normal case—mainly or exclusively symbolic, not ostensive; it does not proceed in virtue of taking any actual or imagined mark to present spatial properties as such, but in virtue of values conventionally assigned to the relevant marks (i.e., words). Moreover, there is a positive injunction to the reasoner who tries to follow such an argument not to draw on any spatial concepts she may possess, on risk of assuming the existence of geometrical properties of objects falling under those concepts that have not been explicitly stated or proved within the argument. It is, precisely, a benefit of symbolic representation that symbols can be chosen whose visual properties do not even roughly resemble the visual properties of the objects they represent, thus removing a possible source of error in reasoning with them.

Such a counterpart argument, in a natural or formal language, may be a valid means to argue to a given geometrical conclusion, but the reasoning involved in following it will be quite different to that involved in following Euclid's argument. So there is, for familiar reasons, no reason to regard it as a satisfactory explanation of what is epistemologically distinctive of the latter reasoning. But the further point that this discussion brings out is the degree to which taking such purely linguistic arguments as a standard or canon can blur fine-grained distinctions between different concepts, and so inhibit analysis of arguments that use diagrams. It is, I suggest, easy for someone who associates concepts with linguistic meanings to think that the definition of a triangle gives *the* concept of triangle *tout court*. I shall be arguing against this specific claim in Chapter 10.

8.4 The Practical Impossibility Objection

The second of Kitcher's objections is the Practical Impossibility Objection. This runs as follows:

> How do we determine that sequences of presentations which we cannot *in practice* achieve are *in principle* possible for us? ... Kant claims that pure intuition can yield the knowledge that line segments are infinitely divisible. Now it is evident that we cannot attain this knowledge by observing a line segment infinitely divided. So what Kant must intend is that we give ourselves a sequence of presentations, showing a continued process of subdivision. Since there are practical limits on our ability to do this, we shall face an awkward question: are these limits reflections of a structural property of experience? To resolve this issue we need, again, that same insight into the structure of experience which pure intuition was supposed to provide.[137]

This objection refers to Proposition I.10 in Euclid, which provides a construction procedure for bisecting a given line. Now on Kitcher's reading of Kant, geometrical truths are 'about some particular feature of the world—that feature in virtue of which they are true'.[138] The neo-Kantian, by contrast, takes Euclid's geometry to be a piece of pure mathematics, as I have mentioned. In relation to the neo-Kantian view, the Practical Impossibility Objection asks whether following Euclid's argument can yield *a priori* knowledge of a general property of finite geometrical lines—here, the property of being infinitely divisible—given that we cannot perform the apparently requisite infinity of acts of construction. We cannot actually draw, and it seems we cannot visualize, the infinite divisibility of a finite geometrical line. So how can drawing diagrams or visualizing figures here contribute to the justification of a geometrical belief?

Questions of infinite divisibility are complex, and some way removed from the argument of Prop. I.32; so it could be readily argued that little in the main line of our discussion hangs on this worry. But the general worry is sufficiently relevant to our wider concerns to deserve consideration here.

In fact, there is a fairly straightforward response to be made: the reasoner can use mathematical induction. Recall that the 'weak' principle of induction on the positive integers states that for any property P, in order to prove that all numbers have P, it is sufficient to prove two things: first, that the number 1 has P; and secondly, that for every positive integer n, if n has

---

[137] Kitcher 1984, p. 51.

[138] Kitcher 1975, p. 29. I discuss Kitcher's argument for this view, and argue that it again misreads Kant, in the Postscript to this chapter.

the property P, then its successor n + 1 also has the property P.[139] Thus, where 'n' ranges over the positive integers, the inference is of the form:

P1
$\forall n\ (Pn \rightarrow (Pn + 1))$

$\rightarrow n\ (Pn)$

Let Pn be defined as the property for any line L of having 2^n parts, where parts have positive length. Then the argument can proceed by induction on the positive integers. The basis step is given by Euclid's construction: Let a visualized line represent L; then the construction given in Prop. I.10 shows that L has 2 (= 2^1) parts. The induction hypothesis is: Let L have 2^n parts. Then the argument runs:

Let *l* be any part of L. Letting the visualized line represent *l*, Euclid's construction shows that *l* has 2 parts.
So each of L's 2^n parts has 2 parts.
But any part of a part of L is a part of L.
So L has 2 x 2^n parts, i.e. 2^{n+1} parts
So, by induction, for any n, L has 2^n parts.
Hence, for any finite line segment L, for any n, L has 2^n parts.[140]

This establishes the desired conclusion, and it does so by a process that involves reasoning with a visualized figure. Note that a reasoner can follow a very similar process in relation to a line drawn on paper. But the conclusion as to infinite divisibility will not relate to the physical composition of the line that has been drawn; it would be irrelevant to object that, at some suitably sub-atomic level, the line might 'run out of parts', so to speak. Rather, the reasoner is using a physical line to represent a geometrical line.

In relation to Prop. I.10, the neo-Kantian claim is that here too—as in the earlier discussion of Prop. I.32—what justifies the general conclusion is first, that this is valid reasoning, which includes reasoning with the diagram, and secondly, that the reasoner's grasp of the relevant construction procedure is sufficient to justify her in believing that the conclusion holds for all such lines. So, provided that these claims as to validity and generality can be made good, the fact that she may not actually be able to bisect the physi-

---

139 On induction generally see any introductory text on set theory or mathematical reasoning; for example, Eccles 1997, Ch. 5.

140 This argument was suggested to me by Marcus Giaquinto.

cal line precisely by using the procedure, nor know whether she has done so, is irrelevant; she can be justified in believing that such a procedure, applied to the geometrical line she takes the diagram to represent, would bisect it. I shall argue for the validity and generality claims in relation to Prop. I.32 in Chapter 10.

I suggest, then, that the neo-Kantian view can readily meet the Practical Impossibility Objection.

8.5 The Exactness Objection

Kitcher's third objection is the Exactness Objection:

> How can we resist the challenge that the presented entities do not have exactly the properties we take them to have?... Just as our powers of ordinary perception are limited and fallible, so too are our powers of mental perception. Because of this we cannot assume that mental perception will give us exact knowledge even of the particular figures we construct. We should concede that we might be unable to distinguish a straight line from one that is very slightly curved. *The concession is dangerous. For imagine that we follow Kant's procedure to arrive at a belief in a geometrical truth. The warranting power of the procedure can be undermined by experiences involving deceptive measurement which seem to show that the statement is only a close approximation to the truth. Given such experiences, it would be rational for us to suppose that our mental visual acuity had failed us, and thus to inhibit formation of the belief.*[141]

We can separate out three distinct questions for the neo-Kantian view from this passage.[142] The first is this: why is the conclusion of Euclid's argument not true just of those (geometrically imperfect) triangles that exactly resemble the diagram actually drawn, and false of geometrical triangles (which do not)? The neo-Kantian should accept that the reasoner's mental powers are finite and fallible, and that a reasoner might in some circumstances be unable to distinguish a straight line from one that is very slightly curved. But even so, she can readily respond to this question by noting the ambiguity in the phrase 'presented entities' in the first line: if 'presented entities' refers to geometrical objects, then she can deny that geometrical objects as represented by figures or diagrams may not have exactly the geometrical properties she takes them to have, for reasons already described. If 'presented entities' refers to the diagrams or figures

[141] Kitcher 1984, p. 51 and p. 53f. Emphasis added.

[142] I am of course again taking the objection as it putatively relates to the neo-Kantian view, not to Kant's own views.

themselves, then the neo-Kantian can again deny that diagrams or figures may not have exactly the physical or visualized properties she takes them to have; for she does not take them to have geometrical properties. As noted, a problem would only arise here if she made the erroneous assumption that she can reason validly by appealing directly to the figure or diagram and trying to 'read off' from it properties of the geometrical object(s) represented.

The second question concerns what conclusion we are supposed to derive from what Kitcher terms the 'dangerous concession': that a reasoner might in some circumstances be unable to distinguish a straight line from one that is very slightly curved. We can imagine situations in which such an inability to discriminate might indeed be dangerous: for example, given a school teacher's request for pupils to measure the internal angles of a diagram of a triangle using a protractor, an approximately correct answer would be highly sensitive to a number of factors: for example, the accuracy of the drawing process, the flatness of the surface, and the pupils' measuring skills. Any deviation from a required norm on these three counts would unsettle the pupils' reasoning, indeed that of any human reasoner, to the desired conclusion (cf. the discussion in Chapter 4). But of course this empirical and inexact process has no bearing on Euclid's argument, to follow which a quite different type of reasoning is required. In the case of Prop. I.32, the reasoning operates by leading the reasoner to grasp certain equalities between angles, without regard to the exact size of those angles. And the reasoning appears to be much more robust than in the empirical case just considered, as a result. Why should this be? In the first place, there is no question of any *measuring* here. Secondly, it seems that even a very inaccurate diagram or figure—one in which lines are bent, overlap or do not meet, for example—can nevertheless be taken by a reasoner to represent a geometrical triangle; and an inaccurate implementation of the construction procedure using such a diagram or figure—line CE drawn slightly but noticeably non-parallel to AB, for example—need not affect the reasoning. The diagram below illustrates this point:

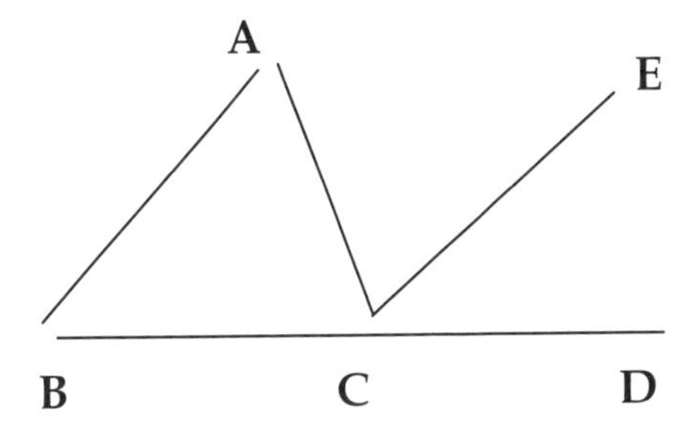

So the final diagram constructed can be visibly inaccurate, without a reasoner being consciously aware of or attending to and correcting the inaccuracy, and yet still be of use for purposes of reasoning.

As yet, then, the concession is not dangerous to the neo-Kantian. What about the case in which the reasoner draws a diagram on what is, without her being consciously aware of it, a curved surface? Isn't there a danger here that she might mistakenly infer that the angles of a triangle sum for example, to more than two right-angles? Note that this is not an unusual or limiting case: this is the situation in which a reasoner who follows Euclid's arguments normally finds herself. But again, we need to bear in mind the crucial distinction between claims about the structure of the physical universe, and claims about space as it is represented in Euclid's geometry. Let us exclude the 'reading-off' view, since it is not a genuine way of following Euclid's argument. Then there seems no reason to think that the hidden curvature of the diagram has any bearing at all on the justification for the reasoner's belief. Recall that the belief does not relate to the physical space occupied by the diagram, but—again—to what the reasoner takes the diagram to represent. And a reasoner who takes the diagram to represent a geometrically curved triangle—a triangle in hyperbolic or elliptical space—has made a mistake. So it does not seem as though the concession is dangerous here either.

What might make the above worry more plausible is that a diagram drawn relatively 'large' on a sphere can have internal angles that are plainly more than two right angles, and perhaps closer or equal to three right angles. This does not seem to be Kitcher's worry: this is not a case of a reasoner's inability to distinguish straight from curved, or of 'deceptive measurement'. But it does suggest that the reasoning here might be rationally defeated by the appearance of the diagram. Hence the third question: is the neo-Kantian view committed to the claim that the justification here cannot be rationally defeated by the visual appearance of the diagram?

The worry is this. Let us imagine hypothetically that justification *can* be defeated by the appearance of the diagram. Then the way seems open for an opponent to claim, by parity of reasoning, that the neo-Kantian has not made out the distinction between her view and 'reading off' views that appeal to the diagram; if she accepts that belief can be defeated by the appearance of the diagram, the claim might go, then she must accept that her positive belief is partly underwritten by the mere appearance of the diagram. But if the neo-Kantian accepts this last claim, then the reasoning must be fallacious. And even if she can avoid this outcome, it would seem hard for the neo-Kantian to retain her view that the reasoning is *a priori* in the face of this challenge.

So the neo-Kantian must reject the possibility of epistemic defeat. But why think this is difficult? In the quotation above Kitcher talks of 'undermining experiences ... which inhibit formation of the belief'. However, we need to distinguish here between two possible inhibitors of belief-formation. *Psychological blockage or hindering* occurs when, in following an argument, a reasoner has a visual experience of the diagram as a result of which she is unable to form (or retain) a given belief, but without having any reason to doubt the (putative) justification of that belief. Perhaps it is just not clear to her how to proceed. *Epistemic blockage or hindering* occurs when, in following an argument, a reasoner has a visual experience of the diagram as a result of which she is unable to form (or retain) a given belief, not because of any psychological difficulty, but because the (putative) justification of that belief is weakened or destroyed by the experience.

Now take the case above, in which a reasoner draws a diagram on a sphere, and let us think of locations on the sphere analogously with the Earth; as having a 'North Pole' at the top as viewed by us, and an 'Equator' running around the middle, etc. With this in mind, consider a reasoner who imagines drawing a diagram of a triangle on such a sphere, with point A on the 'North Pole' and the base lying on the 'Equator' running from point B on 'the Greenwich meridian' to point C at '90 degrees East'. Now imagine that she tries to follow Euclid's argument in relation to such a diagram. It will be straightforward for her to extend the base line BC further around the 'Equator' to a point D (provided she does not go all the way round), as requested by the first step of the construction procedure. But how to draw a line through C parallel to AB? There is no way to do this: a line segment through C would have to be extendable to a great circle parallel to the great circle extending AB. But two great circles cannot be drawn parallel to each other; they must intersect at two points. So the reasoning to line 1 (the claim that $\angle$BAC and $\angle$ACE are alternate) cannot proceed; the reasoner has had a psychologically blocking experience. Such a reasoner can psychologically 'unblock' herself by visualizing a Euclidean plane figure of a triangle, and reason on that basis.

Note that a reasoner's knowledge (or rather, meta-knowledge) that she has been psychologically hindered or blocked in an earlier inference can rationally ground a worry about the epistemic reliability of later inferences; the worry may be that she has failed to take in all the epistemically relevant information as a result. But this is not a case of epistemic defeat, as defined above; it is not a situation in which a visual experience of the diagram is defeating the justification of a given belief. And what the example above brings out is the primacy of the text of the argument over the diagram; it is this that prevents the non-existence of angle ACE in the case above from having epistemic force. If the representational properties of the diagram were not specified by the text, then inconsistencies between the diagram and the text could in principle count against justification. But since the function of the diagram is to represent a situation described in the rubric of the argument, only two possibilities exist: either it does so, in which case the argument as such may proceed; or it does not do so, in which case there is psychological blockage or hindering. In neither case is there epistemic blockage.

8.6 Summary

The goal of this chapter has been to show that we can make room for a positive neo-Kantian view of our target reasoning. The chapter first identified the neo-Kantian view, and distinguished it from Kant's own views. It then reviewed in detail three well-known lines of potential criticism of the neo-Kantian view, beyond the Generality Objection discussed in Chapter 7.

1. The *Irrelevance Objection* suggested that the neo-Kantian could not treat the generalization in Euclid's argument as conceptual, on pain of making the role of intuition irrelevant, and the figure or diagram epistemically unnecessary. However, this criticism misses its intended target, which is not the acquisition of geometrical knowledge in general, but its acquisition using the kind(s) of reasoning required to follow Euclid's argument. Once this further constraint is acknowledged, the claim of epistemic necessity for the diagram (or figure) is very plausible. By distinguishing between spatial and non-spatial concepts, I argued that neo-Kantian view can both treat the generalization as conceptual and retain the idea that there is something intuitive about this reasoning, without conceding that the diagram is irrelevant to it. But this in turn requires a more detailed treatment of the relevant concepts; I turn to this in Chapter 10 below.
2. The *Practical Impossibility Objection* suggested that, in the absence of infinite powers of visualization, the neo-Kantian lacked an

account of the justification of general mathematical claims. In response, I argued that the neo-Kantian can argue for the general claim by using the accepted principle of mathematical induction.

3. Lastly, the *Exactness Objection* raised a series of worries relating to the justification afforded by reasoning using the figure or diagram. The neo-Kantian response was to insist on two distinctions: one between the properties of the representing diagram or figure and the properties of the geometrical entity or entities represented; and one between psychological and epistemic blockage. Once these are understood, Euclid's reasoning was found to be significantly more robust than a parallel but distinct empirical process of reasoning in geometry, consideration of which may erroneously be part of the motivation for these worries.

The effect of this discussion is to start to bring out the character of, and the specific commitments incurred by, what I have termed the neo-Kantian view. In particular, it highlights the extent to which the neo-Kantian view recognizes a spatial aspect to certain geometrical concepts, and an underlying commitment to what might be termed a fine-grained approach to such concepts, within which the spatial/non-spatial distinction can be articulated. It is here, in the specification of concepts, that the deepest contrast is to be found with Kant's own views.

In the next chapter, I turn to consider the logic of Euclid's argument.

8.7 Postscript: The Irrelevance Objection and Kitcher's Kant

In this chapter I have taken various objections originally advanced against Kant by Philip Kitcher in relation, not to Kant himself, but to the neo-Kantian view. In this Postscript I argue that in fact Kitcher seriously misreads Kant, and specifically that the Irrelevance Objection is ill-founded as a result.

As a summary reading of Kant, Kitcher claims that:

> Kant proposes that we construct figures in thought, inspect them with the mind's eye, and thus arrive at *a priori* knowledge of the axioms from which our proofs begin ... Kant's own proposal is tied to a sensuous notion of pure intuition—we draw mental pictures and look at them.[143]

But I suggest that this is misleading, and perhaps mistaken, in three respects. First, as far as I am aware, Kant does not discuss the specific ques-

[143] Kitcher 1984, p. 49-50.

tion of how we know the axioms of Euclid's geometry in any detail in the first Critique, and it is certainly not at the forefront of the discussion in the Doctrine of Method, on which Kitcher's reconstruction is based. As I argued at the end of the last chapter, the focus there is on the epistemology of the reasoning involved in following Euclid's argument, and this does not concern as such—though it does logically presuppose—an account of how the reasoner knows Euclid's axioms. Secondly, Kant's notion of pure intuition is not, I think, sensuous, at least if this term is taken in the normal way as relating to sensations. On the contrary, Kant explicitly dissociates pure intuition from sensation, which he takes as arising in empirical intuition. Pure intuitions are pure in being sensation-free, and they are intuitions in being immediate singular representations. Thirdly, Kant's view is not—as Kitcher later acknowledges[144]—that the reasoner merely draws diagrams and then looks at them. This would be, or would be close to, a reading-off view of the kind rejected (both on its own merits and as an interpretation of Kant) in the previous chapter. Rather, the reasoner follows the argument in relation to the triangles represented by the diagram; and it is this overall process—and not any tacit appeal to specific features of the diagram as such—that ultimately confers justification.

Does Kant's own position succumb to the Irrelevance Objection? I suggested above that it does not, and that there is a strategy available to him that is broadly analogous to that of the neo-Kantian. But there is also a question as to whether the objection, which derives from Kitcher's misreading, ever really gets going against Kant's position. Recall that the Irrelevance Objection presents a dilemma: either the generalization to Euclid's conclusion is conceptual or it is not. If it is conceptual, then—supposedly—it does not depend on intuition, and the figure is redundant. But if it is not conceptual, then the figure must have intuitive properties as well as merely accidental properties, and a non-circular account is required of how a reasoner may differentiate between them.

If Kant is to face this dilemma, then on his account it must be an either/or matter whether the generalization is conceptual or involves intuition. But is this in fact his view? Of course, the disjunction between intuition and concept is fundamental to Kant. But recall that, in relation to the kind(s) of reasoning discussed in the Doctrine of Method, Kant's claim is that the scope of the generalization is determined not by the concept of a triangle as such but by the *schema* of the concept of a triangle; that is, by the reasoner's grasp of what objects are generated by the relevant construc-

---

[144] Kitcher 1984, p. 52. Kitcher 1975 recognizes, if only briefly, the epistemic role of construction procedures in Kant's account (p. 43); but this recognition is not carried forward to Kitcher 1984.

tion procedure.[145] This grasp has a spatial or intuitive component, and schemata are, precisely, for Kant what unite the conceptual and the intuitive.

Kitcher tries to cut off this line of response as follows:

> Geometrical truths must either be about some particular feature of the world—that feature in virtue of which they are true—or they must state some particular property of our concepts. Since they are not analytic, the latter cannot be the case. So geometric truths are true in virtue of some facet of the world.[146]

Call this the 'No-Property' argument. However, in reply Kant can simply deny the move from non-analyticity to the claim that geometrical claims do not state properties of concepts; he can claim that such claims might state properties of concepts understood in relation to their schemata—properties not 'contained in' the concepts—and still be synthetic. And this seems to be Kant's position:[147] compare his remarks in the Schematism,[148] and also (and explicitly) in the Doctrine of Method. In a telling passage from the latter, Kant remarks:

> [In geometrical reasoning] I am not to see what I actually think in my concept of a triangle (this is nothing further than its mere definition), rather I am to go beyond it to properties that do not lie in this concept but still belong to it ... I put together in a pure intuition the manifold that belongs to the schema of a triangle in general and thus to its concept, through which general synthetic propositions must be constructed.[149]

For Kant, this aptly brings out (1) the connection between geometrical concepts and geometrical definitions here, (2) the important contrast between a concept and the schema of a concept, and (3) the explanatory value of the latter. Since geometrical claims rely on schemata for their generality, it does seem that Kant's view is that they are both synthetic and state properties of concepts.

So the 'No-Property' argument fails, and Kitcher's overall argument by dilemma cannot really be formulated fairly against Kant. Of course, I am not suggesting that Kant's view of concepts is not problematic, and there

---

[145] On the role of construction procedures in Kant, see Section 7.4 above.

[146] Kitcher 1975, p. 29.

[147] Cf. Friedman 1992, p. 90.

[148] A141/B180, partly quoted in Section 7.4 above.

[149] A718/B746, emphasis added.

are serious difficulties relating to his conception of a schema, as noted. But this would not affect the point being made here, which is that—at least as formulated—the Irrelevance Objection does not properly arise for Kant, once his position is better understood.

9

The Competing Logics of Prop. I.32

9.1 Introduction

The discussion in Chapters 7 and 8 so far has been designed to show, first, that we can find in Kant a promising account of the kind(s) of reasoning required to follow Euclid's argument in Prop. I.32; and second, that there is room in principle—several well-known objections notwithstanding—to develop a more persuasive neo-Kantian view of this reasoning.

I now want to explore this neo-Kantian view further. In Chapter 6, it will be recalled, I suggested that we should reject the Leibnizian View of our target reasoning. In the course of that discussion, I considered several of the inferences in Euclid's argument, and argued that while some of those inferences were sentential, others were diagrammatic. I also argued that there is no reason to think that diagrams are intrinsically out of place in proofs.

I now want to turn from considering individual inferences to considering Euclid's argument as a whole. Earlier in the discussion I suggested that an interesting distinction could be drawn between what were termed 'naïve' and 'expert' ways of reasoning with the diagram in following Euclid's argument. But how exactly should we understand this distinction? In particular, how if at all can we reconstruct the logic of these different ways of reasoning? And are there other such ways of reasoning to be reconstructed here? The approach I shall take here is a straightforward one. First, I shall review the 'naïve approach/expert approach' distinction, and ask whether we can identify other routes to belief that constitute ways to follow Euclid's argument. Secondly, I shall describe various different types of logical reconstruction, and assess which if any of these are appropriately sensitive to the epistemology of the routes to belief previously identified.

9.2 The Naïve and Expert Approaches Reconsidered

At several points in the discussion so far—notably in Chapter 2—I have suggested that we can detect two distinct ways in which a reasoner can follow Euclid's argument.

1. On the *naïve* approach, the reasoner takes the diagram to have depictive scope; that is, she takes it to represent those geometrical triangles that it visually resembles. (Since visual resemblance is not a precisely specified relation, its representational scope here will include those geometrical triangles that seem to her to be roughly similar to the diagram.) The generalization is then a movement of thought, at the end of the Demonstration, from the narrow claim that the depicted triangles have the angle-sum property, to the general claim that all triangles have that property. This inference is grounded in the reasoner's recognition that no inferential step depends on any property of the triangle(s) depicted that is not a property of all triangles. Note, however, that the scope of the diagram has not changed by the end of the above reasoning. It is a further movement of thought for the reasoner to grasp that, since nothing hinges on the special features of the depicted triangles, the diagram may permissibly be taken to have wide scope from the outset; that is, as representing triangles that it does not visually resemble.
2. On the *expert* approach, the reasoner takes the scope of the diagram to be general from the outset. The final generalizing inference is not a reflective one, for the diagram—and so any claims warranted by reasoning with the diagram—is understood in advance to be general. In this case, it would be wrong to think of the expert reasoner as following the presentation for a particular case, and then making a generalizing inference to the conclusion. Rather, she is reasoning generally throughout.

On both approaches it seems to be true that the reasoner can 'switch off' from her sensitivity to what if anything the diagram represents and 'switch on' again later. That is, the naïve reasoner and the expert reasoner seem alike to be able to (1) attend to the initial diagram, (2) follow the sequence of changes made to the diagram during the course of its construction and the transitions in thought involving the diagram, and then, (3) having reminded herself of what the diagram represents, reason to the general conclusion. There is a certain similarity between this and syntactic reasoning with the symbol-strings of a formalized argument in an interpreted formal logic, where the reasoner may often find it convenient to ignore the semantics altogether, and then later attend to the interpretation of the conclusion derived.

So much is common to the naïve and the expert approaches. However, there is also an interesting phenomenological difference between them. During her experience of following the argument, the naïve reasoner can be quite committed to the thought that she is 'seeing what happens' to a particular diagram, when manipulated in such-and-such a way. It may not seem to her as though she is reasoning about all triangles, or making any universal claims as such. Rather, she is following a particular object of her attention through a series of changes, to its ultimate 'fate'.[150] By contrast, though the expert is reasoning with a given diagram, and so will be sensitive to claims made in the text about the diagram, she will, so to speak, be professionally unconcerned with the identity of the diagram as such; for her the reasoning is general from the outset, and the particularity of the diagram itself is of little or no relevance.

9.3 Is the Expert Approach a Way to Follow Euclid's Argument?

I have argued that both the naïve and the expert approaches constitute valid ways to follow Euclid's argument; neither novice nor expert has gone wrong in following the argument as she does. But someone might nevertheless claim that there is internal evidence to suggest that the expert approach cannot be used to follow Euclid's argument in the way intended. Recall that in Euclid's argument (cf. Section 2.4) there is a clear distinction between the claim made by the initial Enunciation and that made by the subsequent Specification.[151] The Enunciation makes a quite general claim:

> In any triangle, if one of the sides be produced, the exterior angle is equal to the two interior and opposite angles, and the three interior angles of the triangle are equal to two right angles.

The Specification, not surprisingly perhaps, is more specific:

[150] A similar point is made by Kit Fine in Fine 1985, p. 133.

[151] We might note that the Greek term 'diorismos' ('Specification') has the connotation of delineating something, or giving it better definition.

> I say that the exterior angle ACD is equal to the two interior and opposite angles BAC, ABC, and the three interior angles of the triangle ABC, ACB, BAC are equal to two right angles.

How should we understand the logical relationship between these two parts of Euclid's argument?

Note that the Conclusion recapitulates not the Enunciation, but the Specification. That is, the official Conclusion of Euclid's argument is not the real conclusion, if we take the latter to be the general angle sum claim for triangles, following normal practice. So, if the Enunciation is not to be left hanging, the argument must take for granted that the overall claim to be established (given originally in the Enunciation) follows directly from the Conclusion. But what, then, is the intended contrast between the Specification and the Enunciation? If there is to be a genuine contrast here between these two parts of Euclid's argument, then it seems as though the Specification must be intended to make a claim about a particular case, such that a successful argument in this case for that claim is to be taken later, when understood in full generality, as a successful argument for the general proposition. This impression is reinforced by the position of the Setting-Out ('Let ABC be a triangle') between the other two parts, on this reading; for on this reading the Setting-Out invites the reasoner to consider a given triangle and construct a diagram accordingly, and this then provides the particular case required. If this interpretation is right, then it would suggest that Euclid intends this: that the reasoner is invited to establish a particular claim, and then to recognize that the argument for that claim can be reinterpreted as an argument for the general conclusion. But this in turn implies that the expert approach is not available; for on the expert approach there is no transition in thought from particular claim to general conclusion.

Tempting though this line of thought is, it cannot be quite right. If it were, then there would be some specifiable place in the reasoning at which the expert reasoner, who takes the argument to be general from the outset, has gone wrong. But there is no such place, and indeed there is no reason to think that the expert does go wrong. In particular, it is not clear that the term 'ABC' mentioned in the Setting-Out has to be taken as (in effect) a name. On the contrary, there appear in principle to be several different possible logical ways to read the argument, and on some of these readings the term 'ABC' can be taken as a variable. It may be that we can use these latter readings to preserve the expert approach as a means to follow Euclid's argument.

9.4 Are Other Approaches Possible?

Before we consider these logical readings, however, I want to ask whether there may not be other ways to follow Euclid's argument, in addition to the naïve and the expert approaches already considered. We can approach this question by considering different possible sequences of thoughts, as follows. Here 'ASP' refers for convenience to the angle-sum property, and I use 'T' as a placeholder for different types of name, or for a variable, that might figure as a replacement for 'ABC' in the various thoughts. The dots '...' abbreviate the intermediate reasoning, much of which was analyzed in Chapters 2 and 6:

1. Let T be any triangle...T has the ASP. So all triangles have the ASP.

2. Let T be any triangle as depicted...T has the ASP. But the reasoning does not depend on any feature of T not shared by all triangles. So all triangles have the ASP.

(Recall that 'as depicted' here means, in effect, 'of similar shape to the diagram'.) I will take (1) to describe the expert approach, and (2) the naïve approach, as described just above.

The representational scope of the diagram is narrower in (2) than in (1). Are there sequences that constitute ways to follow Euclid's argument, on which the diagram has narrower scope, or none at all? Consider these alternative sequences of thoughts:

3. Let T be the triangle depicted...T has the ASP. But the reasoning does not depend on any feature of T not shared by all triangles. So all triangles have the ASP.

4. Let T be this (pointing to the diagram). This is a triangle...T has the ASP. But the reasoning does not depend on any feature of T not shared by all triangles. So all triangles have the ASP.

We can call (3) the Super-Naïve approach, and (4) the Ultra-Naïve approach. Are these latter approaches genuinely ways to follow Euclid's argument? Consider the Ultra-Naïve approach first. On this approach, the diagram does not represent at all (unless it is in some sense self-depictive), and the role played by 'this' is one of non-deferred ostension; that is, it is supposed to indicate the diagram itself, not some object(s) represented by the diagram. The difficulty with the Ultra-Naïve approach is this: if the reasoner is to carry out the Construction correctly, the diagram must itself instantiate the relevant geometrical properties mentioned. That is, line CE on the diagram must be exactly parallel to line BA. But in following Euclid's argument, the reasoner does not, and is not required to, check or measure between the lines to ensure that they are parallel. So the Ultra-Naïve approach does not describe a genuine way to follow Euclid's argument.

What about the Super-Naïve approach? Here the difficulty lies in the implausibility of the reasoner's commitment to there being one and only one triangle depicted by the diagram: which one is it? Moreover, the reasoner has to deny that the diagram also depicts triangles which may only be very slightly different in visual appearance from that which she takes it to depict; or which are of exactly the same shape but slightly different size; or that are of exactly the same shape but of slightly different orientation. But this is, I suggest, an absurdity to which someone who follows Euclid's argument is not committed. If so, then the Super-Naïve approach is also not a way to follow Euclid's argument.

There remains the possibility that there are other ways to follow Euclid's argument. But this is unlikely, and I shall focus on the expert and naïve approaches in what follows.

9.5 Logical Readings of Prop. I.32

I now want to consider five different candidate logical readings of the reasoning here. The question is which, if any, of them is the best formalization of the particular sequences of thoughts that constitute the naïve and expert approaches. It may be helpful to recall again what these sequences are:

1. Let T be any triangle...T has the ASP. So all triangles have the ASP [expert].

2. Let T be any triangle as depicted...T has the ASP. But the reasoning does not depend on any feature of T not shared by all triangles. So all triangles have the ASP [naïve].

But I also want to leave open a slightly variant pair of sequences:

1*. Let T be a [*not*: any] triangle...T has the ASP. So all triangles have the ASP [expert].

2*. Let T be a [*not*: any] triangle as depicted…T has the ASP. But the reasoning does not depend on any feature of T not shared by all triangles. So all triangles have the ASP [naïve].

The concluding thought in each case is surely the same:

(C) All (Euclidean) triangles have the ASP.

What about the premiss in each case? We can see it as implicitly expressed by the initial 'Let…' sentence, where that is read as an indicative sentence with an optative 'introduction', roughly thus for (1) and (2) above:

(P1) (Let it be the case that) T is any triangle [expert].

(P2) (Let it be the case that) T is any triangle as depicted [naïve].

With this in mind, we can identify five different candidate ways in which one might read either of the premisses above. (Let us assume that the predicate 'F' denotes '… is a triangle', and that 'G' denotes '… has the angle sum property'.) They are:

a) Universal quantifier/bound variable: $\forall x\,(Fx \rightarrow x = a)$
b) Proper name: Fn
c) Existential quantifier/bound variable: $\exists x\,(Fx \wedge x = a)$
d) Arbitrary name: Fa
e) Open sentence: Fx

The Universal Quantifier Reading

I suggest we can rule out (a), the universal quantifier reading, at once in relation to both premisses. On that reading, the thought would be

Everything is such that, if it is a triangle, then it is T.

In other words, T is every triangle. But this dubiously grammatical thought is not the thought expressed by (P1) or (P2), and it cannot be, since it would imply that there is just one triangle, T.

The Proper Name Reading

A second possibility is that the reasoner takes the diagram as, in effect, a proper name denoting a particular geometrical triangle. Kit Fine criticizes this reading as follows:

> [This view] is ... psychologically unrealistic. If we ask ourselves what we are doing ... it is most implausible to suppose that we are reasoning about a particular individual. In this respect, the standard historical example of a triangle is misleading. For in geometric demonstrations of a traditional sort, it is usual to draw a diagram of the figures one is reasoning about; and so there is some room for the hypothesis that one is actually reasoning about those figures. But change the example and the implausibility of the hypothesis becomes immediately clear. Suppose I wish to establish that all natural numbers have a prime factorization. I take an arbitrary natural number n, establish that it has a prime factorization, then conclude that all numbers have a prime factorization. But in this case one is under no temptation to suppose that the reasoning is about a particular number. For which one is it? 3? 15? $10^{10} + 6$?[152]

This argument would be cogent if we had any prior reason to regard the contrasted geometrical and arithmetical reasoning as logically or epistemically equivalent. But Fine gives no reason to think that they are, and there are good reasons to think they are not equivalent. First, the arithmetic example starts from the premiss that the natural number selected is arbitrary. But it is not clear that the premiss thought 'Let ABC be a triangle' is necessarily to be understood *ab initio* by the reasoner as claiming that the triangle is an arbitrary one. Secondly, the arithmetic example does not involve any representation with depictive content. But the diagram is a depictive representation, and it is the contrast between this depictive content and the reasoner's general grasp that in part makes the case an interesting and distinctive one for analysis.

Nonetheless, we should reject the proper name reading. There are two difficulties: first, as noted already (and noted by Fine), it is mysterious how the reasoner can be reasoning about just one individual. Secondly, there does not seem to be any plausible rule of logical inference available from the premisses containing the proper name to the general conclusion. And without that, the reading cannot be used to capture valid sequences of thoughts which do contain such general inferences, such as those in (1) - (2*) above.

---

[152] Fine 1985.

The Existential Quantifier Reading

The existential quantifier reading, advanced by Hintikka, is harder to assess.[153] On the one hand, the treatment of the premiss seems quite natural if we consider the expert and naïve approaches as in (1*) and (2*): for in these it is rendered ‘T is a [*not*: any] triangle ...’, and this is a standard type of case for the use of an existential quantifier. On the other hand, if we prefer (1) and (2) above, then there is something slightly artificial about reading an ‘any’ claim in terms of an existential quantifier.

On this reading the skeleton of the argument can be reconstructed as follows:

$\exists x\ (Fx \wedge x = a)$
Ga

$\wedge x\ (Fx \wedge x = a) \rightarrow Ga$

$\forall y\ [\forall x\ (Fx \ \forall\ x = y)\ \forall\ \ Gy]$

The final inference is one of Universal Generalization, and—as usually presented in formal logic—this is a valid form of inference from a sentence containing an arbitrary name to a general claim, provided that no step in the reasoning relies on any assumption expressed by a sentence containing the arbitrary name. (I will take it that arbitrary names are sufficiently well understood in general to be used in possible readings of Euclid’s argument.)[154]

Again, however, this reconstruction is rather unnatural. For the Universal Generalization to go through, the conclusion must contain a disguised existential antecedent in the conditional. We cannot reject this out of hand; but a more natural reading would be preferable.

---

[153] Hintikka 1967. Hintikka tries to motivate this reading by appealing to a notion of ‘ekthesis’ to be found, he claims, in Euclid and Aristotle. But even if it is granted that the concept of ekthesis is the same in both cases, it is far from clear that it must be read in terms of existential instantiation.

[154] See Lemmon 1965, p. ix and the discussion in pp. 107-9. Fine 1985 contains a detailed discussion of reasoning with arbitrary names.

The Arbitrary Name Reading

The fourth reading uses arbitrary names, but without a prior existential instantiation.[155]

Fa
Ga

Fa → Ga

$\forall$x (Fx $\forall$ Gx)

Again, the final inference is via Universal Generalization.

Now the arbitrary name reading seems to capture the naïve reasoning rather well. In particular, ‘Fa’ strikes an attractive balance as a reading of the premiss. On the one hand, it preserves the desired difference between an ‘any’ claim and an ‘all’ claim, unlike the Universal Quantifier reading; and it is not at odds with reading ‘any’ as ‘a’ in (1*) and (2*) above. On the other hand, it also respects the phenomenology of the naïve reasoner's thinking, as noted earlier: her sense that she is ‘seeing what happens’ to (the object(s) depicted by) it, and the way in which the reasoner follows the diagram through the process of reasoning.

However, the reading has an important drawback: it is far from clear that, on the naïve (or indeed the expert) approach, the reasoner ever entertains a thought with the content ‘Fa $\forall$ Ga’, i.e. ‘If [arbitrary entity] is a triangle, then [arbitrary entity] has the ASP.’ On the contrary, the relevant thought seems to be, again, ‘T [or: arbitrary triangle] has the ASP’ (or perhaps, ‘T is a triangle that has the ASP’); that is, there is no implication apparent here. Moreover, the argument presented here reads Euclid's conclusion as ‘$\forall$x (Fx $\forall$ Gx)’. But, strictly speaking, this formalizes not (C) above, but rather :

(C*) All things are such that: if they are (Euclidean) triangles, then they have the ASP.

---

[155] See the discussion in Beth 1956 and Parsons 1983, Ch. 5 and Postscript, for example.

(C*) is true if and only if all things in the universe have this property: that if they are (Euclidean) triangles, then they have the ASP. This could be vacuously true, given a standard reading of material implication, if there were no (Euclidean) triangles at all. But, though the issues are not clear-cut, it is a standard worry here that this latter condition is not sufficient for the truth of (C). Moreover, it mistakes the target of a reasoner's thoughts. The reasoner here is not reasoning about all things in the universe. She is reasoning about triangles. She is entertaining a thought of the form 'All Fs are G' or similar, and there is no apparent implication here, as the '→' suggests. Hence '∀x (Fx ∀ Gx)' is not an appropriate way to formalize (C).

This argument has broadly tracked one made by Mark Sainsbury, and I shall assume that Sainsbury is correct in claiming that it can be shown that no truth functional sentence connective can be inserted into the '∀ ' position in '∀x (Fx ∀ Gx)' in such a way as to yield an adequate formalization of sentences such as (C).[156] As Sainsbury points out, this suggests that the overall problem lies not in the choice of connective, but in the relation between the universal quantifier and the single open sentence 'Fx ∀ Gx' that it takes to form a sentence; and he uses this analysis to motivate treatment of sentences such as (C) in terms of binary, and not unary, quantifiers. A binary quantifier is a quantifier that takes two open sentences to form a sentence (I use 'A' to denote a binary universal quantifier below; Sainsbury gives a suitable rule of interpretation for the relevant quantifier in Sainsbury 1991).[157]

The appeal to binary quantifiers is not sufficient to settle the worry about vacuous truth-making raised above, as Sainsbury is careful to note. But it does introduce another—and perhaps more natural—means to formalize the overall argument. This is as follows, using arbitrary names:

Fa
Ga

Fa;Ga

Ax (Fx;Gx)

[156] See Sainsbury 1991, Section 4.17 *passim*.

[157] Ibid., p. 197.

This avoids the difficulties mentioned above. Note, too, that the conclusion is more natural than that presented by the Existential Reading. We can also use the device of binary quantifiers for the latter, reconstructing it as follows ('E' denotes an existential binary quantifier):

$$Ay\ [Ex\ (Fx;x = y);Gy]$$

But this does not remove the existential quantifier, and so does not remove the source of the unnaturalness already noted above.

To sum up so far: I suggest that the arbitrary name reading, using binary quantifiers, is both plausible in its own right as a reconstruction of the naïve reasoning, and superior to the alternatives surveyed here.

The Open Sentence Reading

On the open sentence reading, the diagram is understood as, in effect, a variable. Presented using unary quantifiers, the argument is as follows:

Fx
Gx

$Fx \rightarrow Gx$

$\forall x\ (Fx\ \forall\ Gx)$

The open sentence reading faces a well-known objection, summarized by Fine as follows:

> Open sentences ... are not susceptible of truth and falsehood and cannot therefore enter directly into inferential relationships. Since our understanding of variables is tied to their use with variable-binding operators, it must be presumed that the instantial terms are somehow implicitly bound and that the direct object of reasoning is not the open sentences themselves but the closed sentences resulting from their bondage.[158]

On this view, the open sentence reading is only defensible if the sentences are understood as being implicitly bound by a universal quantifier at

[158] Fine 1985, p. 132. Potter 2000 (p. 45) advances what is essentially the same objection in relation to arbitrary names.

each stage of the argument. A thought construed in terms of an open sentence would then be true iff the proposition expressed by the universal closure of the relevant sentence was true; and an inference would be valid iff it was constituted by a movement from one such thought to another such thought.

There are two difficulties with this proposal. First, the initial claim that non-susceptibility of truth and falsehood renders sentences (here, open sentences) incapable of entering inferential relationships is a suspect one. Take an example of Frege's: 'William Tell shot an apple off his son's head.'[159] This proposition is plausibly neither true nor false. Yet it can readily enter into some inferential relationships; we can validly infer that William Tell shot something off his son's head. So we should not accept Fine's first suggestion. Second, we should not accept Fine's claim that 'it must be presumed that the instantial terms are somehow implicitly bound and that the direct object of reasoning is not the open sentences themselves but the closed sentences resulting from their bondage.' The effect of this would be to turn the argument above into the following:

$\forall x\ Fx$
$\forall x\ Gx$

$\forall x\ Fx \rightarrow {\to}\!\!\!\!\!x\ Gx$

${\to}\!\!\!\!\!x\ (Fx \rightarrow Gx)$

But this is plainly invalid, and so not a possible reconstruction of either the naïve or the expert thinking above.

Fine's objection is not successful; and his preferred remedy is hardly satisfactory. We should, I suggest, take the open sentence reading seriously. Again, we can improve the overall form of the argument by using binary quantifiers:

[159] From Frege's 'Logic' of 1897; cf. Beaney 1997, p. 230.

Fx
Gx

(Fx;Gx)

Ax (Fx;Gx)

Presented like this, the reading has much to recommend it. First, the premiss 'Fx' captures the expert reading of 'T is a [or: any] triangle' well. Secondly, the argument above does not contain the suspect open sentence 'Fa → Ga'. Moreover, its conclusion also contains no implicational arrow, and thus offers a more natural reading of (C). Nevertheless, the reading also loses the idea, which was characteristic of the naïve approach, that the reasoner is tracking the changes to, or seeing what happens to, the triangle(s) to which the diagram is similar in shape. If we are to take the variable in the argument above as a *variable*—as we need to, if there is to be a genuine contrast with the arbitrary name reading—then there can be nothing distinctive about the entity or entities to which it refers. But it seems as though we can only understand the thought here as employing a variable if we ignore any tendency the naïve reasoner has to take the diagram as representing triangles to which it is similar in shape.

This makes the open sentence reading unattractive as a construal of the naïve approach. But, by the same token, the reading seems to capture the expert reasoning very well. For what was distinctive about the expert approach was precisely that the expert reasoner took the representational scope of the diagram to be not merely depictive, but entirely general from the outset. And yet at the same time we can retain and make intelligible a distinction between 'all' and 'any' using it—and hold on to the 'any' reading for the premiss, as we cannot using universal quantifiers.

We are now in a position to return to the issue raised in Section 9.3 above. There I identified a question as to whether the expert approach was available at all as a way to follow Euclid's argument. The problem was as follows. Recall that the Enunciation in Euclid makes an entirely general claim:

> In any triangle, if one of the sides be produced, the exterior angle is equal to the two interior and opposite angles, and the three interior angles of the triangle are equal to two right angles.

The Specification, however, is more specific:

> I say that the exterior angle ACD is equal to the two interior and opposite angles BAC, ABC, and the three interior angles of the triangle ABC, ACB, BAC are equal to two right angles.

Euclid's official Conclusion recapitulates the Specification. But the real conclusion, the angle sum claim, is given by the Enunciation. I noted above that the contrast between Specification and Enunciation seems to demand a movement of thought from particular to general, a transition which does not occur on the expert approach.

We can now see a response to this worry: this is that we do not have to take the diagram as playing the role of a name. We can, using the open sentence reading, see it as operating as, in effect, a variable. If we do this, we can agree that there is a transition from the Conclusion of Euclid's argument to what I have termed the real conclusion: the overall claim to be established (i.e. the Enunciation). But we do not need to see this as involving any movement of thought from particular to general. Rather, we can read it as the valid transition from (Fx;Gx) to Ax (Fx;Gx) described above. The expert reading thus remains a valid way to follow Euclid's argument.

9.6 Summary

This chapter has further explored what were termed the naïve and expert approaches to our target reasoning. It identified distinct sequences of thoughts corresponding to these approaches, and surveyed five possible logical construals of these sequences. Of these, I argued that we should reject the first two, using universal quantifiers and proper names. The third, using existential quantifiers, was more plausible; but I suggested that we should prefer a fourth treatment using arbitrary names as a reconstruction of the naïve reasoning, and a fifth using open sentences as a reconstruction of the expert reasoning. If we do this, we can meet a possible worry that the expert approach is not in fact a means to follow Euclid's argument.

10

The Epistemology of Euclid's Argument

10.1 Introduction

I turn now to the last part of my elaboration and defense of a neo-Kantian view of our target reasoning. This view holds, with Kant, that a reasoner can be justified in believing the angle sum claim by following Euclid's argument, that the diagram contributes to that justification, and that the justification is *a priori*. However, it does not postulate the existence of a faculty of intuition. On the neo-Kantian view, to state the matter positively, a reasoner can follow Euclid's argument by using geometrical concepts, together with her visual imagination. I will take it that the existence of a reasoner's general capacity to exercise visual imagination is not in doubt, given the amount of relevant psychological evidence.[160]

Chapter 8 showed that we could make room for a neo-Kantian view in principle, by reviewing and answering some main lines of possible objection. Chapter 9 then analyzed the overall skeleton of the argument, showing that we could formalize the different inferences of the naïve and expert reasoner by using arbitrary names and open sentences.

However, in order to introduce and locate the neo-Kantian view, I have deliberately postponed discussion of four critical epistemological questions. The first directly addresses a key claim made by the neo-Kantian view:

[160] Robertson 2002 usefully lists some of this literature.

1. How can the neo-Kantian view avoid postulating a faculty of intuition?

The second question arises from Chapter 7:

2. Is the reasoning required to follow Euclid's argument valid?

The third and fourth questions arise from Chapter 9;

3. How can the reasoner who follows Euclid's argument be justified in believing that the angle sum claim holds for *all* triangles?
4. If this reasoning is sufficient to justify belief in Euclid's conclusion, is that justification *a priori*?

This chapter addresses each of these questions in turn.

10.2 Spatial Intuition and Spatial Concepts

First, then, let us consider how the neo-Kantian view can avoid the need to postulate a faculty of intuition. It is a characteristically Kantian claim, as I have noted, that the very possibility of our having thoughts with spatial contents depends on the existence of a faculty of intuition. In the case of Euclid's argument, the suggestion is that it is this capacity that enables the reasoner to draw a diagram of a triangle at all ('I construct a triangle by exhibiting an object corresponding to this concept, either ... in pure intuition ... or in empirical intuition' A713/B742). What the neo-Kantian view needs to show is that it can account for the representation of spatial contents in geometrical thoughts, and specifically here for the representation of a triangle in such thoughts, without invoking such a faculty.

In order to make the discussion as specific as possible, I will use the approach developed by Peacocke in Peacocke 1992, which individuates concepts in terms of possession conditions. Although this approach is hardly uncontroversial, it has two particular merits: first, it is relatively familiar, at least to many Anglo-American philosophers, and so I shall do no more than sketch the bare bones of it here; secondly, it is—as far as I am aware—the most developed single philosophical analysis of concepts available. However, nothing is intended to hang on the precise details of Peacocke's approach as such.

On this broadly Fregean view, concepts are constituents of thoughts, and thoughts are the possible contents of various types of mental state. Concepts are not identified with linguistic meanings, though we can use language to describe concepts. Concepts are not truth-evaluable, and do not express propositions, make claims, state facts or convey information. There is thus a strong distinction here between concepts and what might be termed theories: what one might take to be, at a minimum, sets of rationally interrelated thoughts. For the latter are normally the results of reasoning, and do express propositions, make truth-evaluable claims etc. Differentiating theories from concepts allows for the possibility of constitutive explanations of the former in terms of the latter.

On Peacocke's approach, it is assumed that different concepts are individuated by their possession conditions; that is, in terms of sets of necessary and sufficient conditions on what it is for a user to possess the concept in question.[161] The conditions elucidate the concept in terms of what inferences the reasoner who possesses the concept is deemed to find primitively compelling, where an inference is primitively compelling if, broadly, there are no other reasons intervening between premiss and conclusion. Thus in the simplest case, a concept of CONJUNCTION can be specified as that concept possessed by someone who finds inferences of the form 'A, B' to 'A & B' and from 'A & B' to 'A' or 'B' primitively compelling. In the case of perceptual concepts, the possession conditions may mention two types of non-conceptual content. First, they may mention scenario content, where a scenario is a spatial type whose tokens are sets of perceptible features taken in relation to a given origin and set of axes. Second, they may mention proto-propositional or aspectual content; that is, ways of perceiving features such as symmetries, which are too fine-grained for scenario content to capture, but of which it is nevertheless plausible that a perceiver can be non-conceptually aware. Finally, as well as having possession conditions, each concept has a 'determination theory', which is supposed to explain how those conditions, and the world, jointly determine the semantic value of the concept. Thus for the concept CONJUNCTION, the determination theory states that the semantic value of the concept is given by the truth function that makes the inferences mentioned in its possession conditions truth-preserving.

We can say that someone's developing the relevant dispositions to make inferences marks the move to her acquiring a given concept. But these dispositions may not be operative at a given time—I may have the concept NARCISSUS, and see a narcissus without realizing it is one. Moreover, even when they are operative the subject may not be able to articulate or describe what is distinctive of the concept possessed—as, in my own case, with the concept LARCH. Thus when someone possesses a concept, we can identify three distinct aspects or levels of possession: a concept may be possessed *simpliciter* while being inoperative and inarticulable by its possessor; or it may be operative and inarticulable, or it may be articulable. Peacocke's approach applies, in effect, to possession *simpliciter*.

---

[161] See Peacocke 1992, Chapters 1, pp. 6-18. For non-conceptual content, see his Chapter 3.

Let us say, following Peacocke, that a visual concept is one whose possession conditions, correctly stated, mention some feature(s) of the actual or possible visual appearance of objects that fall under the concept. Then a visual concept $TRIANGLE_V$ can be specified, following Peacocke's style of formulation, as follows.

$TRIANGLE_V$ is that concept C for a thinker to possess which:

1. She will believe of any object presented under perceptual-demonstrative mode of presentation m that C(m), whenever the object presented under m occupies a triangular plane region of the scenario that her experience represents as instantiated around her, and she takes her experience at face value; and
2. For an object thought about under some other mode n, she will believe C(n) iff she accepts that the object presented under n has one of the shapes that objects are represented as having by experiences of the kind mentioned in the first clause.

The claim is that these conditions are individually necessary and jointly sufficient to individuate the concept in question. Condition (1) here can mention triangularity in a non-question-begging way, since the concept does not fall within the propositional attitudes of the thinker. Condition (2) accommodates the possibility that someone can believe of an unperceived object that it is visually triangular. Finally, the language of (2) also allows for the fact that something can be one of many different shapes and still be triangular.

The possession conditions for $TRIANGLE_V$ are relatively complicated. But it is more straightforward to give possession conditions for a geometrical concept appropriate to Euclid's definition, for the characteristics mentioned in the definition are, and are fairly clearly intended to be, individually necessary and jointly sufficient for triangularity. Thus we can specify $TRIANGLE_{EG}$ as follows.

$TRIANGLE_{EG}$ is that concept C for a thinker to possess which:

1. If she entertains a thought of an entity m as being a rectilinear trilateral plane figure, then she will judge without further reasons that C(m); and

2. If she entertains a thought of an entity m as C(m), then she will judge without further reasons that m is a rectilinear trilateral plane figure.[162]

TRIANGLE$_V$ and TRIANGLE$_{EG}$ are different concepts. Someone who possessed TRIANGLE$_V$ can in principle draw a triangle ABC, as Euclid requests in the Setting-Out at the start of Prop. I.32. If she also possesses the concept TRIANGLE$_{EG}$, and understands the background conventions as to representation, then she can properly take a diagram thus drawn to depict one or more geometrical triangles of the kind specified in the definition. This is what is required for the reasoner to follow the Setting-Out; parallel remarks apply to the Construction, in which the reasoner is instructed to extend line BC to D, and to draw line CE parallel to line AB.

10.3 Visual Geometrical Concepts

I suggested in Chapter 9 that we could understand the Setting-Out of Euclid's argument as a claim of the form '(Let it be the case that) ABC is a triangle;' that is, as embedding a propositional claim that can serve as an initial premiss for the argument. Differentiating between the visual and geometrical concepts as I did just above allows us to explain how the reasoner is to understand this premiss, given her knowledge of the definition. But we should also note that it is not the case that a reasoner must have Euclid's definition to hand in order to form a suitable geometrical concept of a triangle. On the contrary, she may form such a concept by a process of reflecting on the diagram or figure itself.[163]

How might this occur? Imagine someone who has the visual concept TRIANGLE$_V$ and uses it to draw the following diagram of a triangle:

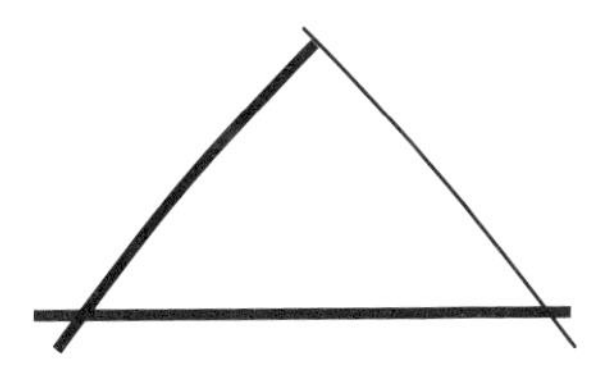

[162] Recall that Euclid's definition of a figure is as 'that which is contained by any boundary or boundaries'; that is to say, figures in Euclid are always 'closed' figures.

[163] I owe this idea, and much else in this chapter, to Giaquinto 1998.

She might well notice that although the two upper sides were straightish, they were not as straight as the bottom line: they could be straighter. And that two of the lines were thicker than a third; they could be thinner. And that while the lines do not quite touch in two places, in a third place one line overruns the other: these lines could touch, and at their points, with no overrun. Now she might imagine being presented with a series of such triangles, in each of which the lines were progressively straighter and thinner, and the angles progressively sharper. At some point, the progression will cross the threshold of her perceptual acuity; she will not be able to distinguish one from another. As far as she can tell, these will be perfect triangles. Does she have to give up any possibility of differentiating later triangles in this succession? No, though she will not normally have any perceptual means to do so. She can still imagine that, though the threshold of her perceptual acuity has been crossed, there are real differences between these subsequent triangles, and so that they can be partially ordered. But she can quite consistently imagine what it would be to be the limit case, and this is a geometrical triangle. She could then take a diagram to represent such a geometrical triangle, without in fact taking the diagram itself to be perfectly triangular. And she is not committed to taking an apparently perfectly triangular diagram to represent an instance of the geometrical concept; she can consistently believe both that the diagram is, as far as she can tell, perfectly triangular and that it is not a representation of a perfect triangle.

This suggests a route via which a reasoner can form a visual geometrical concept, by reflection on an instance of a visual but non-geometrical concept. Take the case of someone who has the visual concept TRIANGLE$_V$, and who is presented with an object that appears perfectly triangular$_V$ to her.[164] We can then specify a concept of perfect triangularity, TRIANGLE$_P$, as follows.

> A concept TRIANGLE$_P$ is that concept C for a thinker to possess which:
>
> 1. For any object n that appears perfectly triangular to her, if she believes m to have the same shape that n appears to have, she will believe without further reasons that C(m); and

[164] Note that 'appears' here should be taken in its core sense of visual appearance. We need to leave open the possibility of other perceptual routes to geometrical concepts. For example, it is plausible that there is a parallel tactual route to a tactual geometrical concept of a triangle. In this case, the possession conditions will mention the tactual 'appearance' or feel of perfect triangularity.

2. If she believes C(m), she will take the apparent shape of an object n to be the real shape of m only if n appears perfectly triangular to her.

Euclid's definition is, then, not strictly speaking necessary for a reasoner to form a geometrical concept of a triangle; she can do by reflection on a visual but non-geometrical concept.

What is the significance of this? In the first place, it suggests that—contrary to an line of thinking advanced in Körner 1960—we should resist the claim that there is a neat partitioning between visual and geometrical concepts: for the concept TRIANGLE_P is both geometrical and visual (note the mention of visual features in the possession conditions). Secondly, the existence of this and similar routes to geometrical concept-formation implies that we are not required to see Euclid's arguments as invalidated by the absence of formally satisfactory definitions. So the possibility that they can be knowledge-yielding is still available. Thirdly, on the independently plausible assumption that the human representation of space is Euclidean, the concepts $\text{TRIANGLE}_\text{EG}$ and TRIANGLE_P are co-extensive. We can now see the utility of Euclid's definition: a reasoner can acquire the non-perceptual geometrical concept $\text{TRIANGLE}_\text{EG}$ directly from the definition, and then move interchangeably between that and the perceptual geometrical concept TRIANGLE_P without introducing any risk of invalid inference, since the truth conditions of thoughts with these conceptual constituents will be unaffected by the switch.

But the main point is, I suggest, this: that this kind of conceptual specification allows us to bring out the precise point of divergence between the neo-Kantian view and the specific set of views attributed in Chapter 7 to Kant. The neo-Kantian view does not invoke a faculty of intuition. But there is nevertheless something intuitive about the concepts we have specified so far. All of these concepts are spatial concepts. We can say that a spatial concept is one whose possession conditions, correctly stated, mention some spatial feature(s) of the objects that fall under the concept. But this is true of all of TRIANGLE_V, $\text{TRIANGLE}_\text{EG}$ and TRIANGLE_P, for in each case the possession conditions mention triangular regions or figures.

In Peacocke's account, a thinker's awareness of environmental features is supposed to be captured in terms of the notion of non-conceptual scenario content. It is a distinct further step to argue for the existence of this type of non-conceptual content, and one to which the neo-Kantian view is not committed. But, though this is more contentious, it is hard to imagine how there could be any genuine explanation of spatial visual concepts (including visual geometrical concepts) that did not mention a thinker's awareness of spatial features of her actual or imagined environment in a non-question-begging way; and if this is true, then we should not think of scenario content as importing special assumptions to which another competing account would not ultimately also be committed. So we can say this: that there is a component of the neo-Kantian account that corresponds to what Kant takes to be a contribution of intuition to geometrical cognition; and that this spatial component is captured here by Peacocke's notion of scenario content.

The neo-Kantian view now has what it needs: an explanation in principle of how a reasoner can entertain geometrical thoughts in virtue of the possession of certain concepts, without explanatory recourse to any postulated faculty of intuition.

10.4 Validity

I turn now to the second question raised above: Is the reasoning required to follow Euclid's argument valid reasoning?

We saw in Chapter 6 that there is no good reason to think that diagrams as such are out of place in proofs; and I argued that the same is true *a fortiori* of arguments more generally. Whether an inference is diagrammatic or sentential has, in and of itself, no bearing on its validity. But now the neo-Kantian view might be thought to face a serious difficulty. Recall again the inference to line 7 of Euclid's argument, which I discussed in Section 6.6:

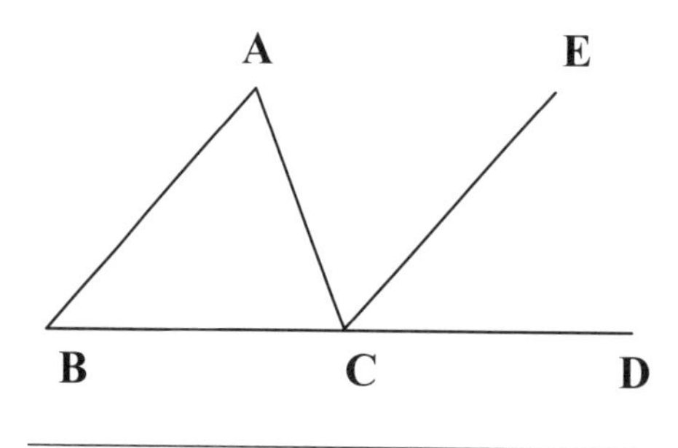

(7) $\angle ACD = \angle ECD + \angle ACE$

I reconstructed this inference in Chapter 2 as follows:

(7a) CE divides ∠ACD into two parts, ∠ECD and ∠ACE, without remainder [from the diagram]

(7b) The whole of an angle is equal in size to the sum of the sizes of any parts into which it is divided without remainder [background assumption]

(7c) ∠ACD = ∠ECD + ∠ACE [7a, 7b: by substitution]

In Section 6.6 I argued that, thus reconstructed, this is a logically valid inference: lines 7a and 7b together entail line 7c (= line 7). But I pointed out that in making this inference the reasoner does not entertain the general thought in line 7b above, and she does not seem to do any substitutional reasoning to reach line 7. This implies that the reasoning used in making the diagrammatic inference is not logical reasoning: there is no logically valid inference from line 7a to line 7c. So the question is how the neo-Kantian can claim that the diagrammatic inference is nevertheless a valid one.

I suggest that the correct response is simply this: the neo-Kantian view can distinguish between inferences that are *logically valid*, and those that are *subject-matter valid*. We can think of an inference from P to Q as subject-matter valid if and only if there is some known truth X of the given subject such that {P, X} together logically entail Q. Since X is a known truth of the subject in question, there can be no question of the inference rule not being truth-preserving. But the rule is not topic-neutral. Its validity depends on the identity of X: it is only because X is a known truth of the subject matter in question that the inference goes through.

With this in mind, we can understand the inference to line 7 above as subject-matter valid, and specifically geometrically valid, thus: though line 7a does not logically entail line 7c, line 7a and the background belief in line 7b geometrically entail it, provided that line 7b is a known truth of geometry. Do we have any reason to doubt that line 7b—the claim that the whole of an angle is equal in size to the sum of the sizes of any parts into which it is divided without remainder—is a known truth of geometry? It is not enunciated in Euclid, as far as I am aware. But surely, we have no reason to doubt it at all.

We can make very similar remarks about other diagrammatic inferences required to follow Euclid's argument—such as the inferences to line 9 and to line 11—as I argued in Section 6.7. For all of these we can say that where an inference is prompted by a reasoner's antecedent geometrical knowledge, we are not required to understand it as logically valid, if valid at all; rather, it may be logically invalid, and yet geometrically valid. As this suggests, restoring the missing premisses is sufficient to give a logically valid reconstruction of the inference, but this comes at the cost of failing to capture what is distinctive of the original thinking.

We can now return to the original question whether the reasoning required to follow Euclid's argument is valid reasoning. Let us review the results of the discussion so far. Chapter 2 identified various types of inference required to follow Euclid's argument at different points. Some of the inferences are clearly logical inferences, and so valid. Others are not logical inferences, but distinctively geometrical inferences, in the sense identified above. In all the cases surveyed, the diagrammatic inferences have obviously valid sentential counterparts. We have seen no reason not to regard these diagrammatic inferences as truth-preserving. To sum up: we need to acknowledge the distinction between logical and subject-matter validity. If we do so, we should conclude that such inferences are valid, in the latter sense. Overall, then, the reasoning required to follow Euclid's argument is valid reasoning.

10.5 Generality

Finally, we can now consider the third question raised above: How can the reasoner who follows Euclid's argument be justified in believing that the angle sum claim holds for *all* (plane) triangles?

Recall that the central problem here is simply that different triangles may have different geometrical properties. In the words of Tennant quoted in Section 6.2:

> One is cautioned, and corrected, about ... the mistake of assuming as given information that is true only of the triangle that one has happened to draw, but which could well be false of other triangles that one might equally well have drawn in its stead.

Tennant's claim elides the distinction between the diagram and the triangle(s) depicted by the diagram: one cannot (normally) draw a geometrical triangle, strictly speaking—though one can draw a diagram that is similar in shape to such a triangle—and I have argued that the claims in Euclid's argument are not to be taken as claims about the diagram, but about the triangles depicted by the diagram. But his point is clear: even if the soundness of an argument as to the depicted triangles could be established, this would be insufficient to establish the truth of the general claim.

We can contrast the position here with a parallel one in the geometry of circles. If we can be justified in believing some claim in (plane) circle geometry on the basis of following a diagrammatic argument for some depicted circle(s), this is sufficient—given that all circles have the same geometrical properties—for us to infer the general claim for all circles. But the crucial 'given' here does not hold for triangles. Since not all triangles have the same geometrical properties, the reasoner needs a justification for generalizing a given valid argument that holds for certain depicted triangles—triangles to which she takes the diagram to be visually similar in appearance—to a claim about all triangles.

I argued in the previous section that the reasoning required to follow Euclid's argument is valid reasoning. So we can at least say this: that Euclid's argument holds for the triangle(s) depicted. That is to say, using the 'arbitrary name' reading adopted in Chapter 9 for illustrative purposes, the reasoner is justified in making the following inference:

Fa
Ga

Fa;Ga

The present question is, using the same reading, how she can make the inference to the overall conclusion:

Fa;Ga

Ax (Fx;Gx)

Note again that, as I argued in Chapter 7, Postscript 2, it is not open to us to understand this generalization as employing an infinite (or indefinitely long) process of iteration; the reasoner does not follow an infinite rule here. But there is a way of putting the question of generality that can seem to make this kind of response obligatory—unattractive though it is—on pain of the question's being otherwise impossible to answer at all. This is to say something like the following: 'How can the reasoner know that *however* she draws the diagram, the argument still holds for the triangles depicted?' The reasoner faced with this question might well conclude that this knowledge is impossible in the absence of an infinite rule: for, even if she knows that the argument holds for some number n drawings of the diagram, how is she to know that it holds for the n + 1th drawing?

However, I suggest we can do better if we approach the question slightly differently, in terms of its converse. The argument will generalize if the reasoner can be justified in believing that no step depends on any property of the triangles depicted that is not a property of all triangles. Can she do this? Let us distinguish between generic and non-generic properties. Generic properties are properties of all triangles (for example, having three sides or having three angles) while non-generic properties (for example, having all three sides equal) are properties only of some subset of all triangles. Then the question is whether the reasoner can be justified in believing that the argument does not depend on any non-generic properties.

At this point we need to recall the construction procedure for Prop. I.32, reconstructed as follows:

Setting-Out:

> Let ABC be a triangle [by Definition 19]

Construction:

> Let BC be produced to D [by Postulate 2]
>
> Let CE be drawn through C parallel to AB, E to lie on the same side of BCD as A [by Prop. I.31]

Each of these claims is warranted by a definition, a Postulate or a previous Proposition. The claim in the Setting-Out clearly does not require or presuppose that the entities represented by the diagram have any particular property and *a fortiori*, it does not presuppose that they have non-generic properties: they must just be triangles. Moreover, it is clear, I suggest, that neither of the two latter claims requires that the triangles represented have any non-generic property. Not only do they not mention any such property explicitly; they are evidently compatible with the diagram's having maximal representational scope. Contrast, for example, a construction procedure that required that the triangles in question be isosceles, or that a given angle should be greater or less than a right-angle. So there is no presupposition in the construction phase of Euclid's argument to the effect that the triangles represented have any non-generic property.

But this is all the reasoner needs to believe in order to make the generalization with justification. Recall that, by the rule of Universal Generalization, if a claim about a given but arbitrary object rests on no prior assumption about the object in question, then it may be generalized into a claim about all such objects. The geometrical analogue of this claim here would be that if a claim about one or more given but arbitrary triangles rests on no prior assumption about the triangles in question, then it may be generalized into a claim about all such triangles. The antecedent of this conditional is satisfied here. No assumption has been made as to the triangles represented by the diagram, and the reasoner who assumes that the diagram is merely depictive has made a mistake. The argument is valid, as we have seen. So the reasoner can make the general claim with justification. Moreover, it is by the distinctively Kantian move of taking the scope of the diagram to be established by the construction procedure, and then exploring the presuppositions of that procedure, that we reach this conclusion.

This is, in effect, the generalization of what I have termed the expert reasoner. But we should note that the same conclusion goes through for the naïve reasoner. In order to be justified in making the generalization, the naïve reasoner must come to recognize on reflection that, though she takes the diagram to be purely depictive, to understand the claims in Euclid's argument as true only of the triangles depicted constitutes a further and unwarranted assumption. By rereading the construction procedure in the light of this recognition, she can be justified in making the concluding generalization.

We can now, finally, diagnose the source of the error in the Rouse Ball fallacy described in Section 1.2. This is that the reasoner mistakenly inferred the faulty claims in lines 17 and 19 from non-generic properties of the objects represented by the diagram. And as described, the source of the error is both clear and correctable by the reasoner.

10.6 Apriority

By following Euclid's argument, then, a reasoner can come to believe the angle sum claim for all triangles with justification. Our principal concern has been with the justification of belief states, as offered by the kinds of reasoning involved here, and in particular by the reasoning with the diagram. So I will not pause to consider the question whether this reasoning is sufficient for knowledge; let me simply note that we have seen no reason so far to doubt that it is.

We can now consider the fourth question raised above: Is this justification *a priori*? In particular, in making the transition from belief states that draw on visual concepts such as $\text{TRIANGLE}_{\text{V}}$ to those that draw on the geometrical concepts $\text{TRIANGLE}_{\text{EG}}$ or $\text{TRIANGLE}_{\text{P}}$, do we not form the latter in reliance on the former? The answer to this is No. There is no reason here to think that the reasoner comes to believe Euclid's angle sum claim by treating her experience of the diagram as evidence and then (mis)generalizing; this is borne out by detailed examination of the various inferences. This is so even on the broad conception of experience described in Chapter 2, which includes not only perception of the external world and the subject's bodily states, but also the subject's awareness of conscious events and states of thinking, imagining and desiring. None of these states is taken here to furnish evidence for the reasoner's belief state. And this belief state can be justified even if she simultaneously holds that her visual experience is not, and could never be, the experience of seeing a geometrical triangle.

Moreover, there are strong general reasons—discussed in Chapters 4 and 5—not to regard the belief-state justification here as *a posteriori*. On the account given here, there can be a causal relationship between the possession of a visual concept and the acquisition of a geometrical concept, but there need be no justificatory relationship between belief states whose contents have these concepts as constituents.

10.7 Is Proposition I.32 a Proof?

The discussion so far naturally raises the question whether Euclid's argument is a proof. Again, I suggest the answer to this is No. Recall that we should ignore the red herring that proofs must be sentential, for reasons discussed in detail in Chapter 6. We can say that an argument is a proof if it satisfies three criteria: it must be surveyable, convincing and rigorous. Euclid's argument in Prop. I.32 is surveyable. It is valid, and I have argued that it is, when correctly understood, convincing. But it is not fully rigorous. It can be given a degree of formalization in a logical language, though not in a way that preserves the distinctive inferences made by the reasoner. Moreover, it is noticeable that proofs of Euclid's conclusions can be rigorously given in a formal diagrammatic system that is intended to be broadly modeled on Euclid's presentation.[165] But the weakness of Euclid's presentation lies in the lack of clarity over the background conventions and assumptions, and over the definitions, as described in Chapter 2. For example: the relevant convention as to the representational scope of the diagram needs to be stated; a suitable continuity assumption must be made explicit; and the status of 'line', 'point' etc. as, in effect, primitive terms needs to be made clear.

What about a similar argument that stated these assumptions, and corrected the definitions so as to be rigorous by modern standards? I suggest this would have a reasonable claim to be considered a proof. But we need not take Euclid's argument to be a proof to regard it as able to justify, or indeed to yield knowledge.

---

[165] Miller 2001.

10.8 Summary

I noted at the outset of this chapter that three critical questions remained unanswered from the discussion so far. These asked how the neo-Kantian view can avoid postulating a faculty of intuition; whether the reasoning required to follow Euclid's argument is valid reasoning; and how the reasoner who follows Euclid's argument can be justified in believing *a priori* that the angle sum claim holds for all triangles. In response, I argued, first, that we can specify different visual and geometrical concepts of a triangle that a reasoner must possess if she is to follow Euclid's argument; and that the reasoner does not require the concept of a triangle given by grasping Euclid's definition in order to follow the argument—she can form a visual geometrical concept from a visual but non-geometrical concept by a process drawing on her visual imagination. These are all spatial concepts, but they do not require us to postulate a faculty of intuition on the reasoner's part. Secondly, I used a reconsideration of certain inferences in the light of the discussion so far to argue that the reasoning required to follow Euclid's argument is valid reasoning. Thirdly, by examining the construction procedure involved, I argued that Prop. I.32 met the standard required for inference by Universal Generalization, so that the reasoner could be justified in believing the angle sum claim generally; and that this justification is *a priori*. Finally, I suggested that though Euclid's argument is not a proof, a similar argument might well be a proof; in any case, the target reasoning was of significant epistemic value.

10.9 Postscript: Analytic Concepts

The reader will recall from Chapter 7 the strong contrast that Kant draws between the fertility of mathematical (here: geometrical) reasoning, and the sterility of philosophical reasoning. I suggest we can use the analysis of concepts in Sections 10.2 and 10.3 above to elucidate Kant's idea here briefly.

In Euclid's argument, I have argued that the diagram can be taken to represent one or more instantiations of the concept $\text{TRIANGLE}_{\text{EG}}$, which the reasoner possesses in view of her understanding of the definition in Euclid of a triangle as a rectilinear trilateral (plane) figure. Kant's point about the sterility of philosophical reasoning is then surely this, that merely analyzing Euclid's definition into its component concepts cannot of itself supply any spatial content; it cannot give a reasoner any of the concepts $\text{TRIANGLE}_{\text{V}}$, $\text{TRIANGLE}_{\text{EG}}$ or $\text{TRIANGLE}_{\text{P}}$. Rather, what makes these concepts spatial—and in particular what allows a reasoner to understand what it is for something to be a *figure*—is derived extra-conceptually.

Moreover, it is only when a reasoner has the relevant spatial concepts—TRIANGLE$_V$, or similar, plus one of the other two geometrical concepts—that she is epistemically in a position to follow Euclid's argument at all. But, as their possession conditions make clear, it is written into the possession conditions for such a concept that the thinker be perceptually or imaginatively aware of the features of her (actual or imagined) environment. Of course, this is hardly sufficient to establish Kant's general claim. But it is nevertheless striking that what confers on a thinker the conceptual capacity to reason by following Euclid's argument is the addition of extra-conceptual spatial content to the concepts that compose Euclid's definition.

11

Conclusions: Value Restored

A young would-be composer wrote to Mozart, asking advice as to how to compose a symphony. Mozart responded that a symphony was a complex and demanding musical form and that it would be better to start with something simpler.

The young man protested, 'But Herr Mozart, you wrote symphonies when you were younger than I am now.'

And Mozart replied, 'Yes; but I never asked how.'[166]

11.1 Introduction

Mozart's reply reminds us that, even in a relatively formal and self-conscious human activity such as musical composition, achievements are often more familiar to us than their explanations. As in music, so—I would suggest—in mathematics. Even the best-known cases of mathematical reasoning can long elude satisfactory explanation. The history of Euclid's argument is a case in point.

11.2 Euclid and the Neo-Kantian View

But it need not be so. In this book I have argued for three claims, as identified in Section 1.4:

166 Solman 2002.

1. *The kind of visual thinking we do in following Euclid's argument can be epistemically valuable—specifically, it can justify belief and confer knowledge.*

In Chapter 2, I described various ways in which a reasoner can follow Euclid's argument, apparently with justification. At certain stages in the argument, there seemed to be two or more different types of inference available. Some used substitution or addition on sentences; others used visual thinking with the diagram, such as a form of 'seeing as', visual translation and visual addition. In Chapter 6, I argued in detail that while it was correct to understand some of the inferences required to follow Euclid's argument as sentential, others should be understood as diagrammatic. In the latter cases, though there might be a counterpart sentential inference, it misdescribed the relevant thinking to construe it in sentential terms.

Following a discussion of the Generality Objection in Chapter 7, Chapter 9 then turned from the nature of the intermediate inferences to the nature of the concluding generalization. It discussed what were earlier termed the naïve and expert approaches to Euclid's argument, the difference between these approaches lying in how the reasoner understands the representational scope of the diagram. I analyzed the overall structure of the argument, and argued that, from a range of standard alternative presentations in a quantified language, we could best understand the naïve and expert thinking overall by reading them in terms of arbitrary names and open sentences, with binary quantifiers, respectively.

Chapter 10 drew a contrast between subject matter (specifically: geometrical) validity and logical validity. It then argued directly for the key epistemic claims: that the reasoning used to follow Euclid's argument was (geometrically) valid reasoning; that reasoner could make the crucial concluding generalization with justification; and that this reasoning was *a priori*. This is not logical reasoning, nor is it sentential reasoning. But it is valid, it is epistemically valuable, and we can use it to come to believe with justification, and I suggest to know, previously unjustified or unknown truths of geometry. In this sense, contrary to Russell's claim in Chapter 1, we can learn geometry from Euclid.

2. *We can identify in Kant an embryonic explanation of such thinking that is preferable to its major alternatives.*

Chapter 3 set up a logically exhaustive Framework of Alternatives, which allowed us to classify candidate explanations of our target reasoning. I argued in Chapters 4 and 5 that we should reject empirical views, on which the experience of the diagram is taken as inductive evidence for the general claim, or on which the reasoning as a whole is taken as deductive but based on inductively established axioms. The Leibnizian View—on which the justification is *a priori* and the reasoning is purely sentential—remains the present orthodoxy amongst philosophers of mathematics. I suggested that a strong version was mistaken in claiming that diagrams were intrinsically out of place in proofs; but that, even without this claim, though it was correct to understand some parts of the thinking required to follow Euclid's argument as using sentential inferences, other parts use diagrammatic inferences. Someone who follows Euclid's argument in the way described in Chapter 2 makes both sentential and diagrammatic inferences. And in the latter, the diagram contributes to justification.

The Kantian View, however, avoided these drawbacks: it did not construe the reasoning as empirical, and it acknowledged the epistemic indispensability of the diagram. So, of the logical alternatives presented in Chapter 3, we should prefer the Kantian View.

There are various attitudes that commentators may in principle strike in relation to the views of a given philosopher, from outright hostility and rejection on one side, to unquestioning endorsement on another. The interpretation advanced here has not been uncritical of Kant: for example, it has ignored discredited Kantian doctrines for Euclid's geometry as the science of space, it does not follow Kant in postulating a special faculty of intuition, and it has argued in detail that Kant's account of construction in pure intuition, and his notion of a schema, are at best unclear.

But on several points I have argued that Kant's position has resources that his critics underestimate; and on the main point, as to the value and nature of the reasoning involved in following Euclid's argument, I argued both that Kant correctly recognizes that the very existence of this kind of reasoning is an important explanandum for the epistemology of mathematics, and that he has a highly promising but so far underexplored explanation of it to offer. Moreover, this explanation is at odds with the very influential present 'logical' tradition of Kant interpretation, which it is tempting to see as structured by what one might term a deep Leibnizianism as to the nature and sources of mathematical justification.

In contrast to other sympathetic critics of Kant, then, the strategy here has been one of direct defense, and not of exculpation or apologia: the claim is not that Kant is elegantly, interestingly or excusably wrong, but that—in these respects at least—he is importantly right.

3. *This account can be developed into a persuasive explanation of the epistemic value of this type of reasoning; one which is recognizably Kantian, but which does not appeal to any special faculty of intuition.*

I have not defended Kant's account as such in any detail here. But I have argued that a development of the Kantian View—what I called the neo-Kantian view—is a persuasive explanation of the epistemic value of our target reasoning. Such an approach faces strong objections, even once it has been separated out from some controversial and specifically Kantian doctrines. But I argued in Chapter 8 that the main lines of objection could be answered, and later chapters have further developed the argument along the lines briefly summarized above. Though the details are quite different, the resulting view of diagrams—on which they are both psychologically and epistemically valuable—bears notable affinities with that of many ancient Greek writers; perhaps this should not be surprising.[167]

It is a feature of the neo-Kantian view that it makes no explanatory appeal to a special faculty of intuition as such; it does not invoke 'intuition' as an independent source of mathematical justification. Rather, justification is conferred in virtue of capacities of the reasoner in whose existence we independently have reason to believe, and which—to some extent at least—we independently understand. Fundamental is the reasoner's capacity to acquire and possess certain visual and spatial concepts, such as those specified in Chapter 10, and to entertain thoughts constituted by those concepts. But we should also note the further capacities made possible by possession of these and related concepts: to take a diagram to represent an object or objects; to follow a set of instructions; to draw or imagine the visual appearance of something; and to understand a sentence as a claim about a state of affairs represented by a diagram.

In by-passing any appeal to a faculty of intuition, has the neo-Kantian given up what makes Kant's account of value? A convincing argument for the existence of such a faculty would be, I take it, a very significant philosophical achievement. But we need not see its absence here as impoverishing: it is simply a different, and more modest, philosophical project to seek to explain the relevant phenomena in terms of capacities in whose existence we already have independent reason to believe. Such a project would fall under a somewhat different tradition in the epistemology of mathematics, which takes as its targets the nature and value of mathematical belief and understanding—and the different ratiocinative processes that warrant mathematical belief and understanding—considered worthy of explanation as naturally occurring phenomena in their own right. This area deserves further investigation. After all, merely to consider the angle sum

---

167 Ancient Greek views on diagrams are summarized and discussed in Knorr 1975, p. 69ff.

claim, there appear to be at least five *other* different diagrammatic routes alone to justified belief.[168]

But the overall point remains: in the right circumstances, there can be a genuine *a priori* epistemology of diagrams in mathematics. And this is, I suggest, a valuable result.

[168] These are: the 'Pythagorean' argument (cf. Heath 1956, p. 320); Thibaut's argument by rotation (ibid., p. 321); an argument by visualization of a triangle as half of two right-angled rectangles (ibid., p. 319); an argument by 'paper-folding' (Roe 1993, p. 1); and Prop. I.32 in Byrne's reconstruction, which involves colored angle segments (Byrne 1847, p. 33).

References

Allwein, G., and Barwise, J. (eds.). 1996. *Logical Reasoning with Diagrams* (Oxford: OUP).

Ayers, M. 1991. *Locke* (London: Routledge).

Barwise, J., and Etchemendy, J. 1998. Computers, Visualization, and the Nature of Reasoning. In T.W. Bynum and J.H. Moor (eds.) *The Digital Phoenix: How Computers are Changing Philosophy* (Oxford: Blackwell).

Beaney, M. 1997. *The Frege Reader* (Oxford: Blackwell).

Berkeley, G. 1988. *Principles of Human Knowledge* (Harmondsworth: Penguin).

Beth, E.W. 1956. Uber Lockes "allgemeines Dreieck". *Kant-Studien* 48.

Blackwell, A. 2001. *Thinking With Diagrams* (Dordrecht: Kluwer).

Boghossian, P. and Peacocke, C. (eds.). 2000. *New Essays on The A Priori* (Oxford: OUP).

BonJour, L. 1998. *In Defense of Pure Reason* (Cambridge: CUP).

Brown, J.R. 1999. *Philosophy of Mathematics* (London: Routledge).

Budd, M. 1989. *Wittgenstein's Philosophy of Psychology* (London: Routledge).

Byrne, O. 1847. *The First Six Books of the Elements of Euclid* (London: Wm. Pickering).

Coffa, A. 1982. Kant, Bolzano and the Emergence of Logicism. *Journal of Philosophy* 79.

Dedekind, R. 1872. Continuity and Irrational Numbers. *Essays on the Theory of Numbers* 1963 (NY: Dover)

Eccles, P.J. 1997. *An Introduction to Mathematical Reasoning* (Cambridge: CUP).

Fine, K. 1985. *Reasoning with Arbitrary Objects* (Oxford: Blackwell).

Forder, H.G. 1927. *The Foundations of Euclidean Geometry* (Cambridge: CUP).

Frege, G. 1885. *Foundations of Arithmetic* (Oxford: Blackwell).

Friedman, M. 1992. *Kant and the Exact Sciences* (Cambridge, MA: Harvard UP).

Friedman, M. 2000. Geometry, Construction and Intuition in Kant and his Successors. Sher and Tieszen 2000.

Giaquinto, M. 1992. Visualizing as a Means of Geometrical Discovery. *Mind and Language*, 7, 381-401.

Giaquinto, M. 1993. Diagrams: Socrates and Meno's Slave. *International Journal of Philosophical Studies*, 1 (1), 81-97.

Giaquinto, M. 1998. Epistemology of the Obvious: A Geometrical Case. *Philosophical Studies* 92, 181-204.

Gillies, D.A. 1982. *Frege, Dedekind and Peano on the Foundations of Arithmetic* (Assen: Van Gorcum).

Glasgow, J., Hari Narayanan, N. and Chandrasekaran, B. 1995. *Diagrammatic Reasoning* (Menlo Park, CA: AIII Press/MIT Press).

Greenberg, M. J. 1993. *Euclidean and Non-Euclidean Geometries* (New York: Freeman).

Greaves, M. 2002. *The Philosophical Status of Diagrams* (Stanford, CA: CSLI Publications).

Hahn, H. 1933. The Crisis in Intuition. In B. McGuinness (ed.) *Hans Hahn. Empiricism, Logic and Mathematics: Philosophical Papers* (Dordrecht: Reidel).

Hallett 1994. Hilbert's Axiomatic Method and the Laws of Thought. In A. George (ed.) *Mathematics and Mind* (Oxford: OUP).

Hammer, E. 1995. *Logic and Visual Information* (Stanford, CA: CSLI Publications).

Harman, G. 1986. *Change in View* (Cambridge, MA: MIT Press).

Harman, G. 1999. *Reasoning, Meaning and Mind* (Oxford: OUP).

Hartshorne, R. 1997. *Companion to Euclid: A Course of Geometry* (Berkeley: Berkeley Mathematics Lecture Notes).

Hartshorne, R. 2000a. *Geometry: Euclid and Beyond* (New York: Springer).

Hartshorne 2000b. Teaching Geometry According to Euclid. *Notices of the American Mathematical Society* 47.

Heath, T. 1956. *The Thirteen Books of Euclid's Elements* (New York: Dover).

Hilbert, D. 1899. *Foundations of Geometry* (La Salle, IL: Open Court).

Hintikka, J. 1967. Kant on the Mathematical Method. *The Monist* 51.

Howell, R. 1973. Intuition, Synthesis and Individuation in the *Critique of Pure Reason*. *Nous* 7.

Jamnik, M. 2001. *Mathematical Reasoning with Diagrams* (Stanford, CA: CSLI Publications).

Jesseph, D.M. 1993. *Berkeley's Philosophy of Mathematics* (Chicago: Chicago UP).

Kant, I. 1998. *Critique of Pure Reason* (Cambridge: CUP).

Kim, J. 1982. The Role of Perception in *A Priori* Knowledge: Some Remarks. *Philosophical Studies*, 40.

Kitcher, P. 1975. Kant and the Foundations of Mathematics. *Philosophical Review* 84.

Kitcher, P. 1984. *The Nature of Mathematical Knowledge* (Oxford: OUP).

Kitcher, P. 2000. *A Priori* Knowledge Revisited. In Boghossian and Peacocke 2000.

Knorr, W.R. 1975. *The Evolution of the Euclidean Elements* (Dordrecht: Reidel).

Körner, S. 1960. *The Philosophy of Mathematics* (New York: Dover).

Leibniz, G.W. 1765/1981. *New Essays on Human Understanding* (Cambridge: CUP).

Lemmon, E.J. 1965. *Beginning Logic* (London: Thomas Nelson and Sons).

Luengo, I. 1996. A Diagrammatic Subsystem of Hilbert's Geometry. In Allwein and Barwise 1996.

MacIntyre, A. 1981. *After Virtue* (London: Duckworth).

Mancosu, P. 1996. *Philosophy of Mathematics and Mathematical Practice in the Seventeenth Century* (Oxford: OUP).

Manders, K. 1995. The Euclidean Diagram. Unpublished paper.

Maxwell, E.A. 1959. *Fallacies in Mathematics* (Cambridge: CUP).

Meserve, B. 1955. *Fundamental Concepts of Geometry* (New York: Dover).

Mill, J.S. 1843. *A System of Logic* (London: Longmans).

Mill, J.S. 1873. *Autobiography* (London).

Miller, A. 2001. *A Diagrammatic Formal System for Euclidean Geometry* (Cornell University: PhD Dissertation).

Mueller, I. 1981. *Philosophy of Mathematics and Deductive Structure in Euclid's Elements* (Cambridge, MA: MIT Press).

Netz, R. 1999. *The Shaping of Deduction in Greek Mathematics* (Cambridge: CUP).

Norman, A.J. 1999. *Diagrammatic Reasoning and Propositional Logic* (University College London: MPhil Thesis).

Peacocke, C. 1992. *A Study of Concepts* (Cambridge, MA: MIT Press).

Parsons, C. 1983. *Mathematics in Philosophy* (Ithaca: Cornell UP).

Prinz, J.J. 2002. *Furnishing the Mind* (Cambridge, MA: MIT Press).

Potter, M. 2000. *Reason's Nearest Kin* (Oxford: OUP).

Potter, R.D. 1999. *Geometric Diagrams and the Problem of Universal Knowledge*, chapter abstracts (Notre Dame: PhD dissertation).

Proclus. 1970. *A Commentary on the First Book of Euclid's Elements* (Princeton, NJ: Princeton UP).

Roberts, D. 1973. *The Existential Graphs of Charles S. Peirce* (The Hague: Mouton).

Robertson, I. 2002. *The Mind's Eye* (London: Bantam Press).

Roe, J. 1993. *Elementary Geometry* (Oxford: OUP).

Ross, D. 1951. *Plato's Theory of Ideas* (Oxford: OUP).

Rouse Ball, W.W. 1905. *Mathematical Recreations and Essays* (London: Macmillan).

Russell, B. 1901. Mathematics and Metaphysicians. Reprinted in *Mysticism and Logic* (London: George Allen and Unwin).

Russell, B. 1903. *The Principles of Mathematics* (London: Routledge)

Russell, B. 1919. *Introduction to Mathematical Philosophy* (London: George Allen and Unwin).

Sainsbury, M. 1991. *Logical Forms* (Oxford: Blackwell).

Sharples, R.W. 1985. *Plato: Meno* (Warminster: Aris and Phillips).

Sher, G. and Tieszen, R. 2000. *Between Logic and Intuition: Essays Presented to Charles Parsons* (Cambridge: CUP).

Shimojima, A. 1996. *On the Efficacy of Representation* (PhD Thesis, Indiana University).

Shin, S-J. 2002. *The Iconic Logic of Peirce's Graphs* (Cambridge, MA: MIT Press).

Shin, S-J. 1994. *The Logical Status of Diagrams* (Cambridge: CUP).

Skorupski, J. 1998. Mill on Language and Logic. In J. Skorupski (ed.) *The Cambridge Companion to Mill* (Cambridge: CUP).
Smit, H. 2000. Kant on Marks and the Immediacy of Intuition. *Philosophical Review* 109.
Solman, J. 2002. *Mozartiana* (New York: Walker and Co.).
Sowa, J. 1999. *Information Representation* (Pacific Grove, CA: Brooks/Cole).
Tennant, N. 1986. The Withering Away of Formal Semantics?. *Mind and Language* 1.
Vlastos, G. 1965. Anamnesis in the Meno. *Dialogue* 4.

Index